Aero and Vibroacoustics of Automotive Turbochargers

Hung Nguyen-Schäfer

Aero and Vibroacoustics of Automotive Turbochargers

Springer

Hung Nguyen-Schäfer
Bosch Mahle Turbo Systems GmbH
Stuttgart
Germany

ISBN 978-3-642-35069-6 ISBN 978-3-642-35070-2 (eBook)
DOI 10.1007/978-3-642-35070-2
Springer Heidelberg New York Dordrecht London

Library of Congress Control Number: 2012955042

Printed on acid-free paper

Springer is part of Springer Science+Business Media (www.springer.com)

Silence is a source of great strength.
(In der Ruhe liegt die starke Kraft.)

Lao Tzu (6th century BCE)

Preface

This book has arisen from my many years of experience in the automotive industry. It is intended for advanced undergraduates in mechanical engineering, research scientists, and practicing engineers working on automotive turbochargers. It could be also used as an aero and vibroacoustics textbook in colleges and universities, and as a practical handbook of Aero and Rotordynamics in turbocharger industries.

Aero and Vibroacoustics of Automotive Turbochargers is a widely interdisciplinary working field dealing with noise source generation and airborne noise propagation from turbochargers to the environment. First, it involves *thermodynamics of turbomachinery* to design the turbocharger and to determine its working conditions. Second, it involves *aerodynamics* and the study of flow dynamics and instabilities of rotating stall and surge in compressors, which lead to aerodynamic noise. These are self-excited vibration noises rather than the response of external excitation noise sources. Third, it involves *fluid and rotordynamics* to study the unbalance whistle due to rotor unbalance, and self-excitation flow instabilities of oil whirls in rotating floating ring bearings, leading to the constant tone (howling). Both noise types are called rotordynamic noise. Lastly, it involves the *noise propagation computation* based on the Lighthill's analogy to investigate the airborne noise of turbochargers in passenger vehicles. However, some mathematical backgrounds in vector, matrix, and tensor analysis are required to comprehend this interdisciplinary working field of Aero and Vibroacoustics.

Airborne noise has recently become a much more important topic in passenger cars next to CO_2 reduction, and reducing fuel consumption. In recent years, engine noise has become relatively quiet so that other noise sources now become audible in passenger vehicles. Drivers and passengers want to use online communications via Internet and mobile phones, or relax with audio and video infotainment media without any external noise disturbance during driving. Therefore, airborne noise level in the cabin is required to be as low as possible. Moreover, car frames made of light metals and light periphery components have been used to reduce the vehicle mass, and to save fuel consumption and reduce emissions. Under these circumstances, induced noise of turbochargers excites the periphery components neighboring the turbocharger, leading to additional airborne noises in the car cabin and environment. The root causes of different airborne noise types in turbochargers, such as pulsation, rotational, growling, whining noise, unbalance whistle, constant tone, crackling noise, high-order harmonic noise, and wear noise are thoroughly studied. Both active and passive measures are recommended to reduce the airborne noise levels in passenger vehicles.

Despite all my careful efforts, there may be some unpredictable errors in this book. I would be very grateful to receive your feedback and hints of errors. Therefore, readers of this book need to have a thorough analysis before applying it to their individual applications, and take their own responsibilities for possible damages.

I am grateful to Dr. Martin Knopf, the executive director of Bosch Mahle Turbo Systems (BMTS) in Stuttgart, for all his support and encouragement. Especially, I am also indebted to my colleague Daren Bolbolan for reviewing this book, and giving me constructive and helpful suggestions. In addition, I would like to acknowledge Dr. Jan-Philip Schmidt at Springer in Heidelberg for proofing mathematical derivations and for editing this book; Mrs. Heather King also at Springer and Dr. Shine David at SPS (P) Ltd. for invaluable recommendations and helps in publishing this book.

Finally, my special thanks go to my wife for her understanding, patience, and endless support during writing this book in my leisure time and on vacation.

Stuttgart, Germany Hung Nguyen-Schäfer

Contents

Abbreviations

Notations

A	Rotor mass, and inertia moment matrix
a	Acceleration
a_v	Bunsen's coefficient
$\mathbf{b}$	External force vector
$\mathbf{C}_{\mathbf{SG}}$	Damping, and gyroscopic matrix
c	Diagonal damping coefficient; radial bearing clearance; absolute velocity; sound speed; damping coefficient
c_c	Cross-coupled damping coefficient
c_m	Meridional velocity component
C'	Compressor characteristic slope
CW	Compressor wheel
c_1	Radial inner bearing clearance
c_2	Radial outer bearing clearance
D_i, D_o	Inner and outer bearing diameter
e	Journal eccentricity
$\mathbf{F}_{\mathrm{B}}$	Bearing force acting on journal
$\mathbf{F}_{\mathrm{J}}$	Journal force acting on bearing
F_1, F_2	Force components in X_1 and X_2
F_r, F_t	Force components in r and t
$\mathbf{f}$	External force vector
h	Oil film thickness
I	Noise intensity
I_p	Polar mass inertia moment
I_t	Transversal mass inertia moment
j	Imaginary unit
$\mathbf{K}_{\mathbf{S}}$	Stiffness coefficient matrix
K	Stiffness coefficient
k	Wavenumber
k_c	Cross-coupled stiffness coefficient

L	Noise pressure level
L_i, L_o	Inner and outer bearing width
L_p	Noise power level
$\mathbf{M}$	Rotor mass matrix
Ma	Mach number
N	Degrees of freedom (DOF)
NVH	Noise vibration hardness
p_0	Ambient absolute pressure
p	Oil film pressure
p'	Noise pressure
P	Noise power
R	Journal radius
RSR	Ring speed ratio
Q	Noise source
r_0	Initial gas bubble volumetric fraction
r	Gas bubble volumetric fraction
s	Rotor eigenvalue
T_0	Ambient temperature
T	Oil film temperature
T'	Throttle characteristic slope; perturbed fluid temperature
t	Time
u	Circumferential velocity
$\mathbf{U}$	Unbalance force and moment matrix
U_r	Rotor circumferential velocity
U_b	Bearing ring circumferential velocity
v	Gas velocity
w	Relative velocity
$\mathbf{x}$	Vibration response

Greek Symbols

α	Attitude angle
β	Blade angle
γ	Journal angular position
ε	Journal relative eccentricity
ζ	Damping ratio
φ	Flow coefficient
λ	Average fluid velocity ratio; eigenvalue
η	Speed ratio
η_{TC}	Overall turbocharger efficiency
η_{liq}	Oil viscosity
η_{mix}	Oil mixture viscosity
η_i	Dynamic viscosity of inner oil film
η_o	Dynamic viscosity of outer oil film

θ	Bearing filling grade
κ	Isentropic exponent of gas
ρ'	Perturbed fluid density
ψ	Pressure-rise coefficient
ω	Whirl frequency
Ω	Rotor angular speed
Ω_R	Bearing-ring angular speed

Chapter 1
Introduction to Turbocharging

1.1 Introduction

Downsized engines, reducing the number of cylinders or volumetric size of cylinders, combined with the use of exhaust turbochargers, are being applied more and more in the automotive industry in order to comply with the recently enacted emission laws for CO_2 and nitrogen oxides (NO_x) reduction in automotive vehicles.

These downsized engines with fewer numbers of cylinders or small cylinder volume induce less friction power between the pistons and cylinders. Additionally, total weight of the vehicle is also reduced due to a smaller engine, leading to less driving friction forces. Both measures help to reduce emissions and improve the specific power of small engines [1]. The fuel consumption of passenger vehicles could be reduced by nearly 18 % without loss of engine power through the combined use of a turbocharger (TC), high-pressure direct injection (HPDI), low-pressure exhaust gas recirculation (LP-EGR), variable compression ratio (VCR), and variable valve train (VVT). With extreme engine downsizing, using turbochargers and hybrid techniques, the reduction of fuel consumption could reach 30 % in total. Such measures could help the automotive industry reduce the enacted average emission law for CO_2, reduced from 120 g/km in 2012 to 95 g/km in 2020 for new passenger vehicles; and from 175 g/km in 2014 to 147 g/km in 2020 for new light duty commercial vehicles in the European Union (EU) countries, as shown in Figs. 1.1 and 1.2.

After the combustion of fuel in the engine, the large energy of exhaust gas still remains in the form of enthalpy from the engine outlet because the temperature of exhaust gas is quite high (e.g., diesel 820 – 850 °C and gasoline 950 – 1050 °C). Generally, this enthalpy energy escapes from the engine to the environment. Why do we not capture it to boost the engine for improving its specific power? To do that, an exhaust-gas turbocharger is necessary to collect the exhaust gas in the turbine, and compress the intake air in the compressor to a high pressure for the engine combustion. This procedure is called turbocharging, which improves

H. Nguyen-Schäfer, *Aero and Vibroacoustics of Automotive Turbochargers*,
DOI: 10.1007/978-3-642-35070-2_1,

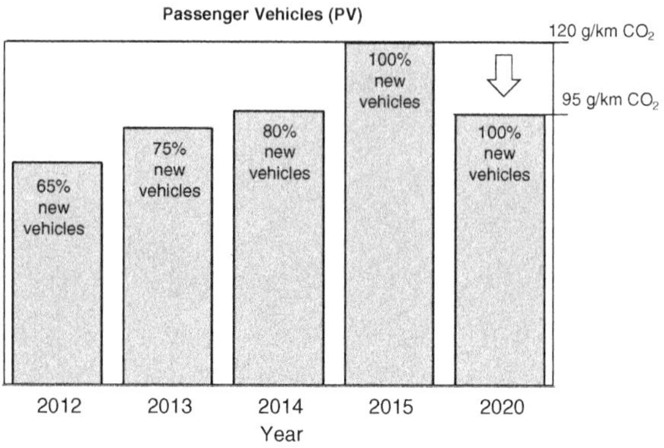

Fig. 1.1 CO_2 average emission limits for new passenger vehicles

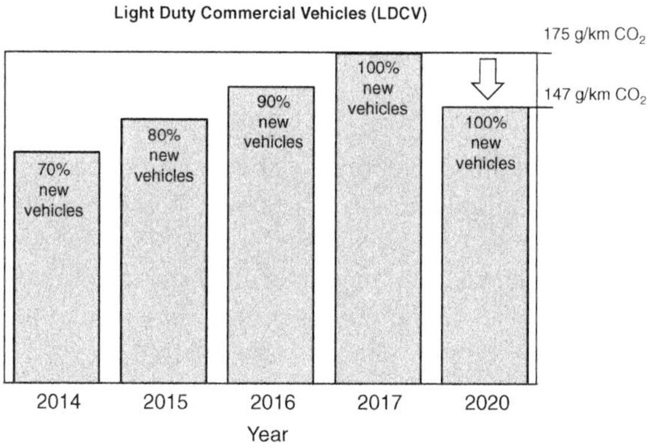

Fig. 1.2 CO_2 average emission limits for new commercial vehicles (LD)

the engine-specific power. The exhaust-gas turbocharger consists of the CHRA (center housing and rotating assembly), as shown in Fig. 1.3, turbine, compressor, and actuator [2]. Both compressor and turbine wheels are mounted and fixed in the rotor shaft that is supported on the bearing system including two separate rotating floating radial bearings, and the thrust bearing. The exhaust gas expands in the turbine wheel and generates the rotational kinetic energy for the rotor shaft. The created turbine energy drives the rotor shaft and compresses the charge air to a high pressure by means of the compressor.

The concept of using exhaust-gas turbochargers takes advantages of improving specific engine power and as well as reducing CO_2 and NO_x. Enabling either having the same power of the original engine but less fuel consumption, which reduces carbon dioxide and as well as nitrogen oxides in the downsized engine or achieving more power at the same fuel consumption in the turbocharged engine without downsizing for those who need more driving fun. However, the first one is an important key to help us comply with the new emission law for the automotive vehicles (see Sect. 1.2).

In this chapter, automotive turbochargers are applied to downsized engines to comply with the given reduction of CO_2 emission from 2012. Due to car

Fig. 1.3 Automotive turbocharger's CHRA (Courtesy BMTS)

infotainment and online communications via Internet while driving, induced airborne noise in the passenger cabin is required to be kept as low as possible, especially in passenger vehicles. Additionally, light car frames and peripheral components are used to reduce total vehicle mass; in turn, to save fuel consumption (mpg), leading to reduction of exhaust gas emissions. Under such boundary conditions, noise induced by the engine and turbocharger is transmitted through the turbocharger, air filter, air intercooler, exhaust gas system (catalyzer, diesel particle filter, and muffler), and car frame into the passenger cabin and car environment. This transmitted noise is called airborne noise that should be strongly reduced in passenger vehicles.

Besides engine noise, some synchronous and asynchronous (super- and subsynchronous) noise types induced in automotive turbochargers are caused by the flow instabilities (rotating stall and surge) in the compressor wheel, rotation of compressor and turbine wheels, rotor unbalance, oil whirls in hydrodynamic radial bearings, and outer race clearance in rolling element bearings. The airflow noise is defined as the *aerodynamic noise* (aeroacoustics). The unbalance whistle, constant tone in the oil-film radial bearings, and high-order harmonic noise in the rolling element bearings are called the *rotordynamic noise* (vibroacoustics or rotordynamic acoustics). Due to nonlinearity of the automotive turbocharger's bearing characteristics, non-synchronous rotor responses are induced from superimposing the synchronous rotor unbalance excitation (1X) on subsynchronous components of the nonlinear rotor system, such as inner and outer oil whirl vibrations [2, 3]. In turn, the inner oil whirl generates subsynchronous noise that is usually called "constant tone", a self-excited vibration response. Necessary measures must be applied to turbochargers in order to minimize these turbocharger noises in passenger vehicles. For this reason, aerodynamics in radial compressors and nonlinear rotordynamic responses of the rotor have been thoroughly studied in a passenger vehicle's turbocharger running on two separate rotating floating ring bearings. Computational results and measurements give a hint that inner oil whirl is the root cause of constant tone. The induced noise of the turbocharger excites other neighboring components nearby and is transmitted to the vehicle cabin and environment.

1.2 Applying Turbochargers to Downsized Engines

In automotive applications, single-stage turbochargers are normally used at the air compression ratio to nearly 2.5. There are two layouts of turbochargers with exhaust gas recirculation (EGR) at low pressure (LP-EGR) and high pressure (HP-EGR). By means of engine exhaust gas recirculation, the burning temperature spike of exhaust gas is lowered, leading to reducing NO_x. Generally, turbocharged downsized engines with the same original power could save nearly 10 % fuel consumption by reducing the cylinder volume by 25 %. To abide by the current emission law or to boost engines of passenger vehicles, two-stage turbochargers are applied to engines working at large charge-air compression ratios above 4. Two-stage turbochargers consist of two single-stage turbochargers with different sizes that are sequentially set up in the engine. Some applications of different turbocharger types are discussed in the following subsections.

1.2.1 Single-Stage Turbochargers with EGR

Figure 1.4 shows a single-stage turbocharger with HP-EGR in which the exhaust gas partly returns to the engine with a maximum of 50 % EGR rate before entering the turbine. The EGR rate is defined by the ratio of the recirculation mass flow rate to the outflow mass flow rate of the exhaust gas. The EGR valve regulates the EGR mass flow rate so that NO_x emission fulfills the current emission law. Due to the expansion energy of the exhaust gas, the turbine (T) propels the compressor (C) that compresses the intake air to the maximum pressure ratio of nearly 2.5. The compression process in the compressor brings the inlet air from the ambient condition to high pressure and temperature as well. To maintain the compressed

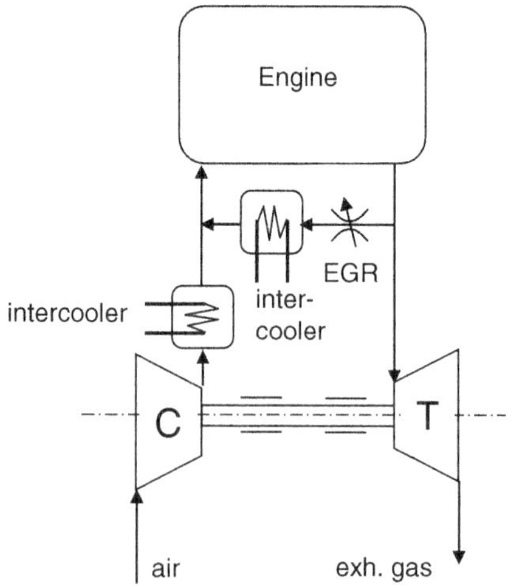

Fig. 1.4 A single-stage turbocharger with HP-EGR

Fig. 1.5 Single-stage turbochargers with LP-EGR

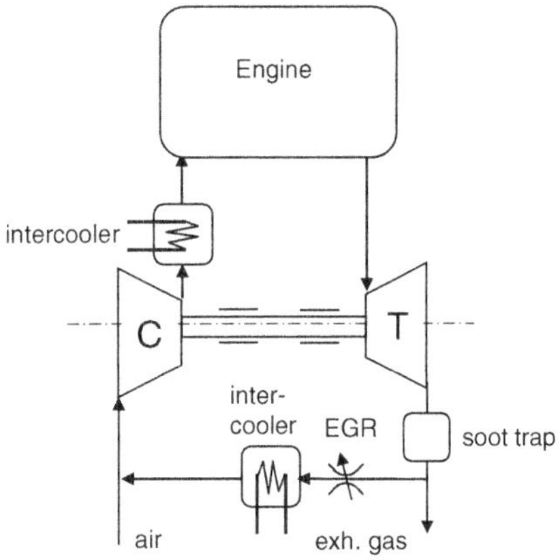

charge air with large mass flow rate for the engine combustion, the intercooler cools down the charge air temperature after the compressor. Otherwise, the air mass flow rate would be reduced due to low charge air density at higher temperatures. In case of turbochargers with HP-EGR, the recirculated exhaust gas pressure after the intercooler must be higher than the charge air pressure at the engine inlet. Hence, the pressure ratio of the turbine is required high enough to overcome the charge air pressure. Generally, the turbine with HP-EGR is designed to be smaller than the turbine without HP-EGR so that the exhaust gas pressure remains higher than the compressed charge air pressure. Therefore, the engine must work against the high exhaust gas pressure at the engine outlet. As a result, the fuel consumption increases compared to the turbocharging system without HP-EGR, especially at high turbocharging pressures. That is the reason why the HP-EGR valve should be closed as soon as the exhaust gas meets the requirements of the current emission law. To overcome this disadvantage, the HP-EGR valve is replaced in the low-pressure site at the turbine outlet. The turbocharger layout shown in Fig. 1.5 is called turbocharger with LP-EGR. Its advantage is that the engine working condition against the high exhaust gas pressure drops; therefore, the fuel performance becomes much more efficient.

However, the LP-EGR turbocharger has two disadvantages: first, the low exhaust gas pressure of the turbine outlet and pressure drop in the soot trap limit the LP-EGR mass flow rate, hence the EGR rate; second, the exhaust gas and ambient air mix together at the compressor inlet; they are compressed in the compressor to higher pressure. Due to the pressure increase in the compressor wheel, the exhaust gas condenses, leading to erosion on the surface of the wheel. Additionally, the unburned hard particles in the exhaust gas impact and damage the aluminum compressor wheel at high rotor speeds, especially the blades at the inlet of the compressor wheel. To prevent damage to the compressor wheel, nickel phosphor (NiP) coating layer with HV from 600 to 1,000 is usually applied to the compressor wheel. However, this causes a reduction in the mass flow rate of the charge air, and therefore the compressor power as well. Due to low-cycle fatigue (LCF) of the driving cycle, the compressor wheel lifetime is reduced.

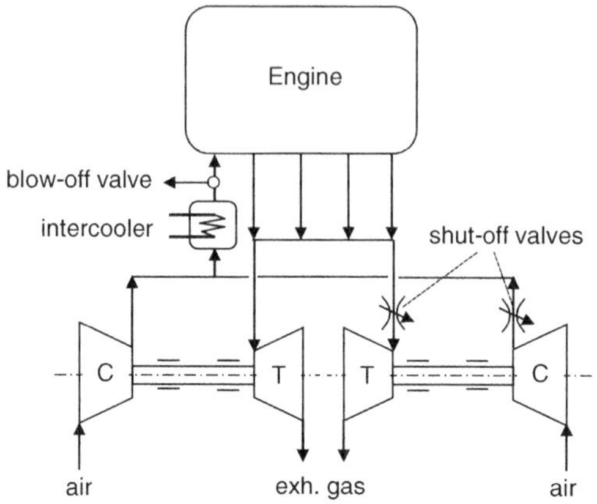

Fig. 1.6 Bi-/Twin-Turbochargers with using intercooler

1.2.2 Biturbochargers

Bi-/Twin-turbochargers consist of two small turbochargers of the same volumetric size; they are parallel setups and operate at the same time in the entire rotor speed range (parallel bi-/twin-turbo). The mass flow rate of exhaust gas is divided into both turbines of the bi-turbo at opening both shut-off valves. In case of an engine with four cylinders, two cylinders provide exhaust gas for each turbocharger, as shown in Fig. 1.6. After compression and cooling, both compressors provide the engine with the total compressed charge air.

With small geometry of turbochargers, the rotor generally has small mass inertia moment that speeds up quickly to achieve maximum torque; therefore, the transient behavior of the parallel bi-turbo is strongly improved in the low-end torque (LET). However, a small turbocharger delivers too little air mass flow rate alone and does not fulfill the required nominal engine power. Hence, both small turbochargers have to operate in parallel in order to deliver enough compressed charge air for the required nominal engine power.

In case of the sequential bi-turbo that is a parallel setup and sequentially operating, only one of the bi-turbo, e.g., the left one operates alone at the low engine speed by closing both shut-off valves. The small turbocharger builds up the charge air pressure much earlier and improves the transient behavior of the turbocharger at low engine speeds. At high engine speeds, the other turbocharger (i.e. the right one) is additionally turned on by opening both shut-off valves where both turbochargers work parallel like the parallel bi-turbo. Hence, the mass flow rate of the charge air increases to empower the engine at high rotor speeds. In strongly powered W-engines with 16 cylinders (e.g. Bugatti Veyron 16.4 Super Sport with a power of nearly 1,200 hp), four single turbochargers (sequential quad-turbo) are used; every turbocharger is applied to four cylinders (4 × 4).

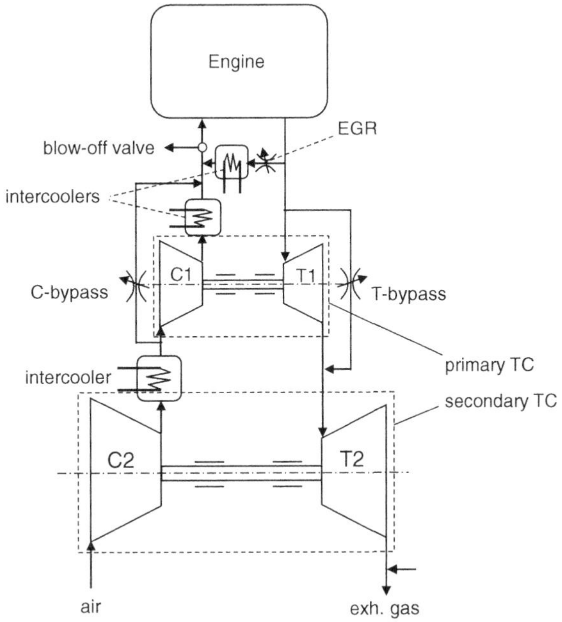

Fig. 1.7 Regulated two-stage turbochargers with HP-EGR

1.2.3 Two-Stage Turbochargers

Two-stage turbochargers are used for high-pressure ratios in which two different volumetric sizes of turbochargers are sequentially set up and work at the given procedure regulated by the bypass valves (see Fig. 1.7). Leaving the engine, the exhaust gas containing large enthalpy is at high pressure and temperature; therefore, the turbocharger at the primary stage (C1, T1) is smaller than the secondary stage (C2, T2). In case of unregulated pressures in sequentially working condition for both stages in commercial vehicle applications, the primary turbocharger is normally about 15 % smaller than the single-stage turbocharger; the secondary turbocharger is 15 % larger than the single-stage one.

At low engine speeds, only the small turbocharger works alone because of its small mass inertia moment; therefore, the transient behavior is significantly improved compared to the large one. During this time, the bypass valves of the primary turbocharger are closed, and the waste gate of the secondary turbine is fully opened so that the secondary turbocharger is nearly decoupled from the two-stage turbocharger. At middle engine speeds from about 1,500 rpm, both turbochargers work sequentially by means of the regulated bypass valves. As soon as the required charge air pressure is reached at high engine speeds, only the secondary turbocharger works alone because the larger turbine (T2) has large efficiency at high rotor speeds. In this case, the primary turbocharger will be decoupled from the two-stage turbocharger, in which both bypass valves of the primary turbocharger are fully opened.

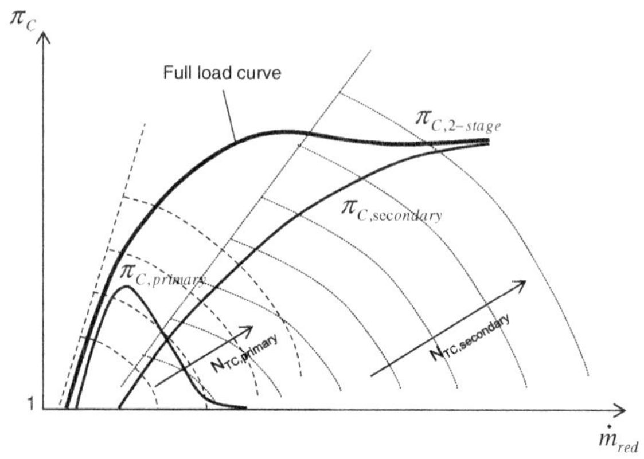

Fig. 1.8 Compression pressure ratios of regulated two-stage turbochargers

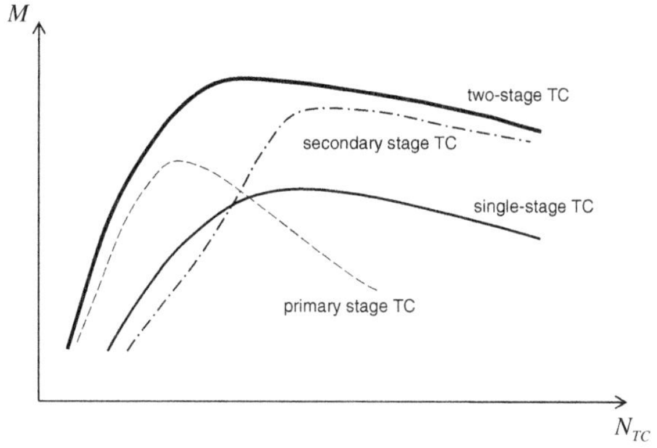

Fig. 1.9 Engine torques of single- and two-stage turbochargers

Similar to the earlier turbochargers, the compressed charge air temperature must be cooled down by air intercoolers before entering the engine in order to maintain the high mass flow rate of the intake air for the required nominal engine power.

The compressor pressure ratio in the full load curve of the two-stage turbocharger is displayed in Fig. 1.8. This pressure ratio results from the product of the pressure ratios of two compressors of the two-stage turbocharger. Due to the small volumetric size of the primary turbocharger, its pressure load curve speeds up faster at low engine speeds to achieve a good transient response compared to the transient response of the secondary turbocharger in LET.

The engine torques of the single- and two-stage turbochargers versus the rotor speed of turbocharger are compared to each other, as illustrated in Fig. 1.9. The single-stage turbocharger is larger than the primary but smaller than the secondary stage of the two-stage turbocharger. Therefore, the transient response of the two-stage turbocharger is better than the single-stage turbocharger. The advantages of the two-stage turbocharger are the better response behavior, higher compression ratio of the charge air, and higher engine torque in the operating speed range.

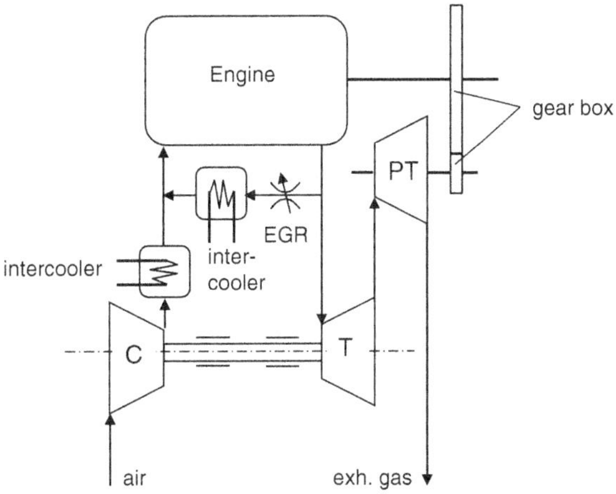

Fig. 1.10 Turbo compounds with a power turbine (*PT*)

1.2.4 Turbo Compounds

Turbo compounds consist of a turbocharger and power turbine (PT) that uses the exhaust gas from the turbocharger, still at high temperature, to generate additional engine power for the engine. The power turbine shaft is directly geared to the engine drive shaft to increase the engine torque and power to help to increase the engine efficiency, as shown in Fig. 1.10. Turbo compounds are possibly applied in commercial vehicles to gain some additional driven energy to the engine shaft.

References

1. Golloch, R.: Downsizing bei Verbrennungsmotoren (in German), Springer, Berlin (2005)
2. Nguyen-Schäfer, H.: Rotordynamics of Automotive Turbochargers. Springer, Berlin (2012)
3. Nguyen-Schäfer, H., Kleinschmidt, R.: Analysis and nonlinear rotordynamics computation of constant tone in automotive turbochargers. 17th ATK Conference, Dresden (2012)

Chapter 2
Induced Noise Types

Noises, as shown in Fig. 2.1 induced in automotive turbochargers are normally classified into the following different types of noises [1, 2].

Pulsation whistle is caused by slight differences in the compressor wheel chamber volumes due to milling or molding process variations. Its frequency ranges from 1,200 to 4,500 Hz and occurs at engine speeds from 1,500 to 3,500 rpm, usually in second gear with high load. It is also dependent of the number of these unequal compressor wheel chambers.

Rotational noise (rotating-blade-related noise) is generated by the rotating blades of the compressor or turbine wheels at an engine speed range from 1,400 to 2,500 rpm in second and third gears. This noise has very high frequencies between about 8 and 18 kHz or higher, which result from the number of blades and rotor speed. Although adult human ears cannot recognize noise higher than 16 kHz, animals in the car could hear such high frequencies.

Growling noise (compressor-stall-related noise) is induced by the partial reversal of the charge air in the compressor wheel. Partial flow separation at the suction side near the blade outlet causes the growling noise. Its frequency ranges in a wide band between 1,200 and 3,500 Hz containing partly metallic noise, and occurs at engine speeds between 1,400 and 2,500 rpm in second and third gears. Growling noise propagates in the direction of the compressor outlet, and the charge-air intercoolers.

Whining noise (compressor-surge-related noise) is induced by a deep surge condition in the compressor wheel where the charge air totally recirculates from the compressor outlet to compressor inlet. The whining noise occurs when driver suddenly releases the gas pedal, causing the required charge-air mass flow rate for the engine to be suddenly reduced at still high turbocharger speeds. This leads to the deep surge working condition in the compressor. Its frequency ranges in a wide band from nearly 800 to 2,700 Hz containing partly metallic noise, and occurs at engine speeds from 1,400 to 2,500 rpm in second and third gears. Whining noise propagates in the direction of the compressor inlet and the air filter system.

Unbalance whistle is caused by the unbalanced rotor and unbalance change of the rotor after a long-term operating time. Its harmonic frequency (1X) is between

H. Nguyen-Schäfer, *Aero and Vibroacoustics of Automotive Turbochargers*,
DOI: 10.1007/978-3-642-35070-2_2, © Springer-Verlag Berlin Heidelberg 2013

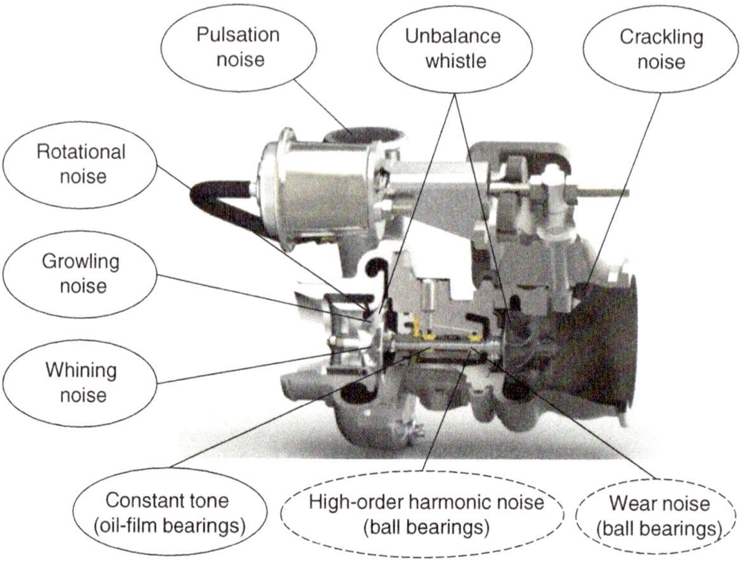

Fig. 2.1 Induced noises of automotive turbochargers

1,200 and 4,500 Hz happening at an engine speed range between 1,500 and 3,500 rpm in second gear at relatively low loads.

Constant tone (howling) is induced by the inner oil whirl due to oil whirling in the oil-film radial bearing; its frequency lies between 600 and 1,000 Hz in the human audible range. The inner oil whirl frequency order in the rotating floating ring bearings reduces from about 0.4 to 0.3X as the rotor speed increases. Therefore, the inner oil-whirl induced frequency varies in a small range between 600 and 1,000 Hz, and is generally considered as quasi-constant in the automotive turbocharger operating speed range. Constant tone often occurs at engine speeds between 1,500 and 3,500 rpm in second to fifth gears with middle to high loads.

High-order harmonic noise often occurs in turbochargers using rolling element bearings (ball and roller bearings) where the constant tone fortunately does not happen. This noise type is caused by an inappropriate design of the oil damper clearance between the bearing outer race or cartridge and bearing housing [1]. The high-order harmonic noise has multiple harmonic frequency orders of 2, 3, 4, 5X, and above, as well as modulation sideband frequencies.

Wear noise mostly occurs in turbochargers using rolling element bearings (ball and roller bearings) if wear defects of balls, rollers, cage, inner, and outer races take place. The wear noise has different asynchronous frequencies that will be discussed in Sect. 4.5.

Crackling noise mostly takes place in wastegated turbochargers in which the waste gate (WG) vibrates about its pivot axis. This vibration generates the metallic crackling noise in various asynchronous frequencies between about 7 and 15 kHz in the WG open position. However, the crackling noise strongly reduces or diminishes when the WG is in the closed position due to the damped vibration amplitude at the waste gate seat located in the turbine housing.

The first four noise types are caused by aerodynamics in the compressor and turbine wheels. The unbalance whistle is induced by the rotor unbalance at high rotor speeds, and the constant tone is generated in the inner bearing hydrodynamic oil film. Both unbalance whistle and constant tone are therefore called rotordynamic noises. Rotordynamics engineers have studied both unbalanced whistle and constant tone and found that an appropriate two-plane rotor trim balancing at high balancing speeds reduces the unbalance whistle but not the constant tone; however, the constant tone could be released at an extremely large rotor unbalance.

In this book, the root causes of these noise types are studied; measures of reducing them are thoroughly discussed. Furthermore, efforts are also made to physically understand the noise behaviors and to reduce the undesirable noise generated in automotive turbochargers. Aerodynamic flows in compressor wheels are used to figure out the causes of the aerodynamic noises of the pulsation whistle, rotational, growling, and whining noises. Rotordynamics is applied to analyze the rotordynamic noises of the unbalance whistle, constant tone, crackling noise, high-order harmonic noise, and wear noise. And finally, some measures are discussed to improve the noise behavior (NVH) of automotive turbochargers in passenger vehicles.

Figure 2.2 shows the noise sources and response noises in a typical automotive turbocharger. The aerodynamic noise sources "1" of the pulsation, rotational, growling, and whining noises are transmitted from the turbocharger through the compressor housing to the air filter, air supply pipes, and air intercooler that are excited by the aerodynamic noise, leading to the response noises. The whining noise generally excites the intake-air supply components due to the flow reversal from the compressor outlet to the compressor inlet. On the contrary, the growling noise affects more the charge-air components, such as the charge-air duct and intercooler, as displayed in Fig. 2.2. The partial flow separation near the

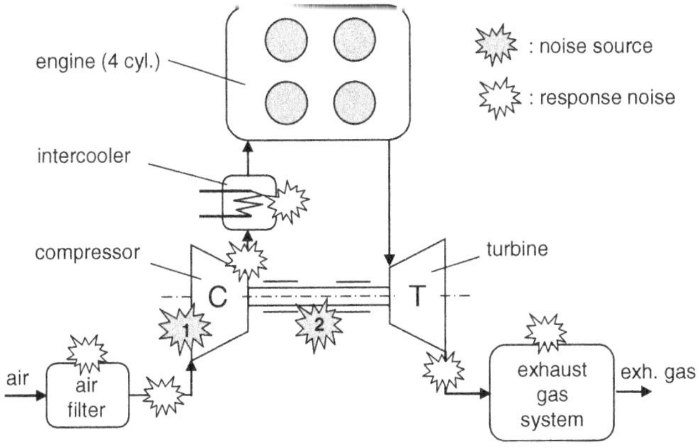

Fig. 2.2 Airborne noises of automotive turbochargers

compressor outlet is to blame for the growling noise propagating direction in the charge-air components; this will be discussed in detail in Sect. 4.1.

The rotordynamic noise sources "2" of the unbalance whistle, crackling noise, and constant tone in oil film bearings/high-order harmonic or wear noise in ball bearings are propagated through the bearing oil films/bearing balls and oil damper, and bearing center housing to the exhaust-gas system of the catalyzer, diesel particle filter (DPF), and muffler. The noise sources also excite the components neighboring the turbocharger, leading to the additional response noises. All induced noises are eventually transmitted through the car frame as airborne noises in the car cabin and environment. The audible airborne noise level depends on the vehicle types and operating engine speeds. Therefore, the delivery and limit acceleration unbalance levels must be determined for each car type; this will be discussed later in Sect. 4.2.

References

1. Nguyen-Schäfer, H.: Rotordynamics of Automotive Turbochargers. Springer, Berlin (2012)
2. Nguyen-Schäfer H., Kleinschmidt, R.: Analysis and nonlinear rotordynamics computation of constant tone in automotive turbochargers. 17th ATK Conference, Dresden (2012)

Chapter 3
Acoustic Propagation Theory

3.1 Aeroacoustic Characteristics

Aerodynamic noise types such as pulsation, rotational, growling, and whining noise are generated from airflows in the compressor of turbocharger. Rotordynamic noises are the noise types like unbalance whistle, constant tone, crackling noise, and high-order harmonic noise. The unbalance whistle is caused by rotor unbalance; the constant tone is induced by the inner oil whirl occurring in the oil-film radial bearings due to self-excitation instability. Additionally, the high-order harmonic noise and possible wear noise mostly occur in the rolling element bearings. Induced noise is transmitted through the bearing oil films, bearing center housing, compressor housing, air filter system, charge-air intercoolers, exhaust-gas manifold, exhaust-gas system (catalytic converter, particle diesel filter (DPF), and muffler), and car frame to the cabin, as shown in Fig. 3.1. The induced noise excites the bearing center housing and periphery components near the turbocharger, such as air filter, charge-air intercoolers, and exhaust-gas system of catalyzer, DPF, and muffler. The excited vibration responses emit airborne noise into the turbocharger's environment. Such airborne noise is uncomfortable and undesirable for the vehicle occupants, and should be reduced as much as possible in passenger-type vehicles.

Figure 3.2 shows some wave characteristics of acoustic propagation, such as vibration amplitude $\xi(t)$, oscillating air-particle velocity $v(t)$, perturbed noise pressure $p'(t)$, and propagating wave front with a sound speed c. At first, an air-particle oscillates with an angular frequency ω and impacts on the neighboring air particles in a compressible fluid (e.g. air). The impact process continues onto the next air particles, leading to propagating waves in the air. This process is compared to the domino effect where the first domino falls with the velocity v; it affects the second one that further affects the next one, and so on. During falling of the domino pieces, the audible noise propagates in the air at the speed of sound c, which is caused by the noise pressure perturbation. In this case, air particles are accelerated and decelerated. This causes a compression wave with $v > 0$ at increasing air pressure and density; an expansion wave with $v < 0$ at decreasing air

H. Nguyen-Schäfer, *Aero and Vibroacoustics of Automotive Turbochargers*,
DOI: 10.1007/978-3-642-35070-2_3,

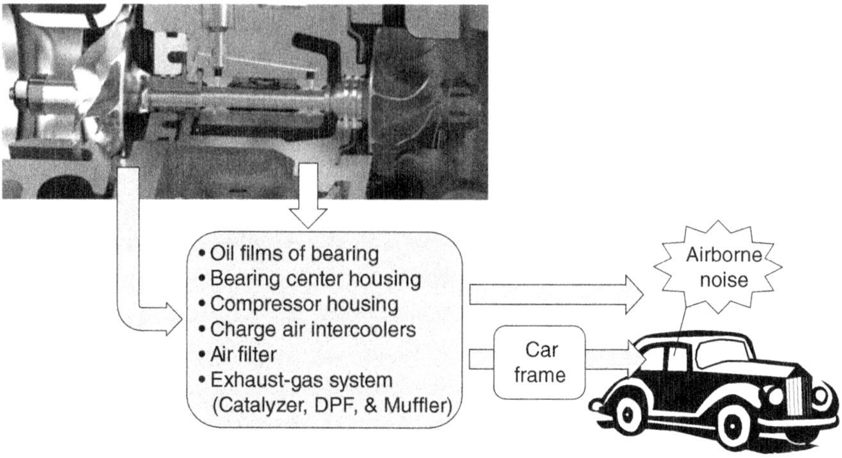

Fig. 3.1 Transmitting ways of airborne noise

pressure and density in a period of time *T*, as shown in Fig. 3.2. At $v = 0$, the oscillating air particle reaches the maximum amplitude where the air pressure and density equal the time-averaged or undisturbed air pressure and density (p_0, ρ_0).

The compression and expansion waves take turns changing in the wavelength λ in the propagation direction. After the compression wave, the expansion wave follows in a period of time *T* and vice versa. While the air particles oscillate with the velocity *v(t)* about their equilibriums, the perturbed noise pressure p' moves with the sound velocity *c* in all propagation directions. However, both velocities *v* and *c* are in the same direction of the wave propagation direction.

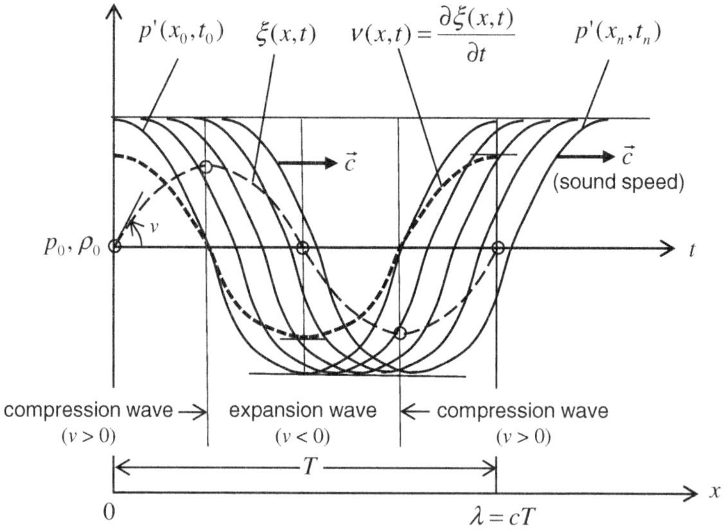

Fig. 3.2 Acoustic propagation wave characteristics

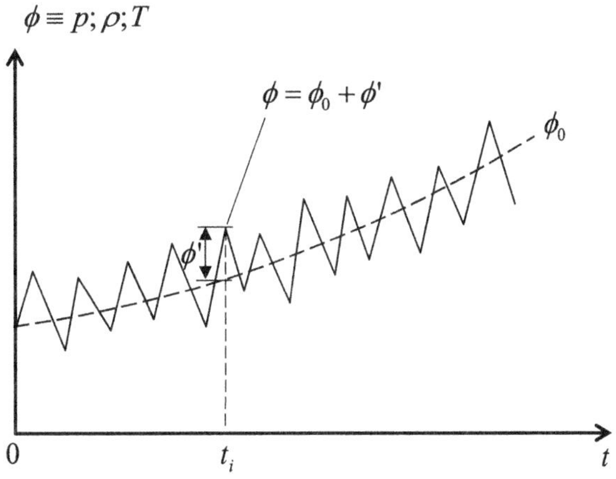

Fig. 3.3 Noise wave characteristics at a given position x

Some thermodynamic characteristics of noise transmitting in a compressible fluid are necessary for noise pressure propagation in the linear aeroacoustics (see Appendix A). The linear aeroacoustics is based on the linearized Euler's equations (LEE) for inviscid and isentropic flows in which the acoustic propagation characteristics of p', ρ', and T' have relatively small perturbation amplitudes in the fluid.

(a) Speed of sound

The sound velocity c is generated from molecular interactions between the oscillating air particles with the fluctuation velocity v about the equilibriums and only depends on the absolute fluid temperature of the transmitting medium. The speed of sound c is defined by the square root of the change rate of the thermodynamic absolute fluid pressure p to its density ρ in an isentropic process.

$$c = \sqrt{\left(\frac{\partial p}{\partial \rho}\right)_s} \tag{3.1a}$$

The acoustic characteristics p', ρ', and T' are defined as (see Fig. 3.3).

$$\begin{aligned} p &= p_0 + p'(x,t); \\ \rho &= \rho_0 + \rho'(x,t); \\ T &= T_0 + T'(x,t) \end{aligned} \tag{3.1b}$$

where

p_0	is the time-averaged or undisturbed gas pressure ($v = 0$);
$p'(x, t)$	is the perturbed noise pressure ($p' \ll p_0$); for convenience only, it is further called the *noise pressure* p';
ρ_0	is the time-averaged or undisturbed gas density ($v = 0$);
$\rho'(x, t)$	is the perturbed gas density ($\rho' \ll \rho_0$);
T_0	is the time-averaged or undisturbed gas temperature ($v = 0$);
$T'(x, t)$	is the perturbed gas temperature ($T' \ll T_0$) thus, $T \approx T_0$.

The relation of pressure and density of a perfect gas is described in an isentropic process.

$$p = C\rho^{\kappa} \tag{3.2}$$

where
C is an integral constant;
κ is the isentropic coefficient of gas ($\kappa_{air} = 1.4$).

The state equation of a perfect gas gives the relationship between pressure, gas density, and temperature.

$$p = \rho RT \tag{3.3}$$

where
R is the gas constant;
ρ is the gas density;
T is the absolute gas temperature [K].

Having substituted Eqs. (3.2 and 3.3) into Eq. (3.1a), one obtains the sound speed in a compressible fluid of a perfect gas (e.g. air) at $T \approx T_0$.

$$c = \sqrt{\frac{\kappa p}{\rho}} = \sqrt{\kappa RT} \approx \sqrt{\kappa RT_0} = c_0 \tag{3.4}$$

Equation (3.4) gives the sound speed in dry air $c_0 \approx 340\ \text{ms}^{-1}$ at $T_0 = 293.14$ K (20 °C).

The longitudinal sound speed in incompressible media (e.g. solid materials and liquids) is calculated by [2]

$$c_s = \sqrt{\frac{E(1-v)}{\rho(1+v)(1-2v)}};\ c_l = \sqrt{\frac{K}{\rho}} = \sqrt{\frac{E}{3\rho(1-2v)}} \tag{3.5}$$

within
E is the medium elasticity modulus (Young's modulus);
v is the medium Poisson's ratio ($v = 0.25$ to 0.30 for most metals);
K is the medium bulk modulus;
ρ is the medium density.

As a result, the sound speed c_s is nearly 6,000 m/s in steel: c_l is about 1,500 m/s in liquid water at 25 °C according to Eq. (3.5).

The relation between the noise pressure p', perturbed gas density ρ', and perturbed gas temperature T' is derived from Eq. (3.1b) and the state equation of a perfect gas, as given in Eq. (3.3).

$$\frac{p}{p_0} = \frac{\rho}{\rho_0} \cdot \frac{T}{T_0} \Leftrightarrow \left(1 + \frac{p'}{p_0}\right) = \left(1 + \frac{\rho'}{\rho_0}\right) \cdot \left(1 + \frac{T'}{T_0}\right)$$
$$\Leftrightarrow 1 + \frac{p'}{p_0} \approx 1 + \frac{\rho'}{\rho_0} + \frac{T'}{T_0} \Rightarrow \frac{p'}{p_0} = \frac{\rho'}{\rho_0} + \frac{T'}{T_0} \tag{3.6}$$

Using Eq. (3.2), it can be written for a perfect gas at $p' \ll p_0$ and $\rho' \ll \rho_0$ in an isentropic process.

$$1 + \left(\frac{p'}{p_0}\right) = \left(1 + \frac{\rho'}{\rho_0}\right)^{\kappa} \approx 1 + \kappa \left(\frac{\rho'}{\rho_0}\right)$$
$$\Rightarrow \left(\frac{p'}{p_0}\right) \approx \kappa \left(\frac{\rho'}{\rho_0}\right) \tag{3.7a}$$

Therefore, one obtains the relation between the perturbation noise pressure and density according to Eqs. (3.4 and 3.7a).

$$\frac{p'}{\rho'} \approx \kappa \left(\frac{p_0}{\rho_0}\right) = \kappa R T_0 = c_0^2 \tag{3.7b}$$

Thus, the perturbed noise pressure is proportional to the perturbed density at very small perturbation amplitudes with $c_0 \approx c$ in linear aeroacoustics.

$$p' = c_0^2 \rho' \approx c^2 \rho' \tag{3.8}$$

The perturbed gas temperature is calculated from Eqs (3.6 and 3.7a).

$$\frac{T'}{T_0} = \frac{p'}{p_0} - \frac{\rho'}{\rho_0}$$
$$= (\kappa - 1)\frac{\rho'}{\rho_0} \approx \left(\frac{\kappa - 1}{\kappa}\right)\frac{p'}{p_0} \tag{3.9}$$

According to Eq. (3.9), the perturbed air temperature $T' \approx 0.57.10^{-4}\ T_0$ and perturbed air density $\rho' \approx 1.43.10^{-4}\ \rho_0$ at a perturbed noise pressure $p' = 20$ N/m^2 and $p_0 = 10^5$ N/m^2. This result shows that $T' \ll T_0$ and $\rho' \ll \rho_0$, as given in Eq. (3.1b).

(b) Noise impedance

The impulse equation of an air particle oscillating with an amplitude ξ at a velocity v is written in the direction x.

$$\rho_0 \frac{\partial^2 \xi}{\partial t^2} = -\frac{\partial p'}{\partial x} \tag{3.10}$$

Thus,

$$\frac{\partial^2 \xi}{\partial t^2} = \frac{\partial v}{\partial t} = -\frac{1}{\rho_0} \cdot \frac{\partial f(t - x/c)}{\partial x} \tag{3.11}$$

The forward propagating noise pressure can be written in the direction x.

$$p'(x,t) = f(t - x/c) \equiv f(\tau)$$

within the delayed time is defined by

$$\tau \equiv (t - x/c).$$

Having timely integrated both sides of Eq. (3.11), one obtains the propagating noise velocity v using the differentiation chain rule.

$$\begin{aligned} v &= -\frac{1}{\rho_0}\int \frac{\partial f(\tau)}{\partial \tau}\cdot\frac{\partial \tau}{\partial x}dt = \frac{1}{\rho_0 c}\int \frac{\partial f(\tau)}{\partial t}\cdot\frac{\partial t}{\partial \tau}dt \\ &= \frac{1}{\rho_0 c}\int \frac{\partial p'(x,t)}{\partial t}\cdot\frac{\partial t}{\partial \tau}dt = \frac{1}{\rho_0 c}\int \frac{\partial p'(x,t)}{\partial t}dt \\ &= \frac{p'}{\rho_0 c} \end{aligned} \tag{3.12}$$

where

$$\frac{\partial \tau}{\partial x} = -\frac{1}{c};\ \frac{\partial \tau}{\partial t} = 1$$

Noise impedance Z is defined by the ratio between the noise pressure and air particle velocity; it is resulted from Eq. (3.12).

$$Z \equiv \rho_0 c = \frac{p'(x,t)}{v(x,t)} \tag{3.13}$$

According to Eq. (3.13), the noise impedance ($\rho_0 c$) in air is about 428 kg/(m^{2}s) at 0 °C (273.14 K); approximately 415 kg/(m^{2}s) at 20 °C (293.14 K). Thus, the air particle velocity v is approximately 0.05 m/s at a noise pressure $p' = 20$ N/m^2 (Pa).

(c) Noise intensity

Noise intensity vector **I** is the product of the noise pressure and air particle velocity.

$$\mathbf{I}(x,t) = p'(x,t)\,v(x,t) \tag{3.14a}$$

where

$p'(x,t)$ is the noise pressure, as given in Eq. (3.1b);
$v(x,t)$ is the air particle velocity.

The time-averaged noise intensity vector is given over the time period T.

$$\mathbf{I}_{\text{mean}}(x) = \frac{1}{T}\int_0^T p'(x,t)v(x,t)dt \tag{3.14b}$$

The rms noise intensity amplitude is calculated from Eqs. (3.13 and 3.14a).

$$\mathrm{I}_{\mathrm{rms}}(x) = \frac{p'^2_{\mathrm{rms}}(x)}{Z} = \frac{p'^2_{\mathrm{rms}}(x)}{\rho_0 c} \tag{3.15}$$

where

p'_{rms} is the root mean square of the noise pressure

$$p'_{\mathrm{rms}}(x) = \sqrt{\frac{1}{T}\int_0^T p'^2(x,t)dt} \tag{3.16}$$

(d) Noise power

Noise power is defined by the sum of the noise intensity over the surface S surrounding the noise source.

$$P = \oint_S \mathbf{I}.\mathbf{n}\, dS \tag{3.17}$$

The time-averaged noise power is resulted from the time-averaged noise intensity and surrounding surface.

$$\bar{P} = \bar{I}S = \frac{S}{T}\int_0^T \mathbf{I}(t) \cdot \mathbf{n}\, dt \tag{3.18}$$

(e) Noise levels

There are two kinds of noise levels in the aeroacoustics: noise power level and pressure level.

- Noise power level

Noise power level L_W [dB] is defined by

$$L_W = 10 \log_{10}\left(\frac{\bar{P}}{P_{\mathrm{ref}}}\right) \tag{3.19}$$

within

$\bar{P}$ is the time-averaged noise power [W], as given in Eq. (3.19);

P_{ref} is the reference noise power (=10^{-12} W).

The time-averaged noise power is proportional to the noise intensity according to Eqs. (3.15 and 3.18); therefore, one can write in

$$\overline{P} \propto I_{\text{rms}} \propto p'^2_{\text{rms}} \tag{3.20}$$

- Noise pressure level

Noise pressure level L_P [dB] is defined by

$$L_P = 10\ \log_{10}\left(\frac{I_{\text{rms}}}{I_{\text{ref}}}\right) = 10\ \log_{10}\left(\frac{p'_{\text{rms}}}{p_{\text{ref}}}\right)^2 = 20\ \log_{10}\left(\frac{p'_{\text{rms}}}{p_{\text{ref}}}\right) \tag{3.21a}$$

where

p'_{rms} is the rms noise pressure [N/m^2], as given in Eq. (3.16);
p_{ref} is the reference noise pressure (= 2.10^{-5} N/m^2).

According to Eq. (3.21a), the noise pressure level is about 114 dB at an effective pressure $p'_{\text{rms}} = 10$ N/m^2 (Pa); $L_P = 120$ dB at $p'_{\text{rms}} = 20$ N/m^2 (Pa). As a result, the noise pressure level increases by nearly 6 dB if the rms noise pressure is doubled. Obviously, the difference in noise pressure level depends on the rms noise pressure ratios.

$$\Delta L_P = L_{P,2} - L_{P,1} = 20\ \log_{10}\left(\frac{p'_{2,\text{rms}}}{p'_{1,\text{rms}}}\right) \tag{3.21b}$$

3.2 Acoustic Propagation Equations

Vibrating solid surface causes local propagating disturbances, such as timely change rates of pressure, density, and temperature in the fluid at the solid surface. The gaseous fluid molecules interact with each other and radiate noise from the noise source to the transmitting medium. Note that the larger the timely change rate of the gas density or pressure, the more effective the noise propagates with radiation velocity **v** that is very close to the average molecular gas velocity in the gas medium [1]. Acoustic propagation is caused by the fluid molecular interaction due to the initial disturbances at the interface between the solid surface and fluid. The noise propagation is based on the continuity and momentum equations in the gaseous medium that is assumed as a perfect gas (e.g. air).

The conservation of mass (continuity) equation of gas with the absolute gas density ρ is written in

$$\begin{aligned} &\frac{\partial \rho}{\partial t} + \nabla.(\rho \mathbf{v}) = 0 \\ \Leftrightarrow\ &\frac{\partial(\rho_0 + \rho')}{\partial t} + \nabla.(\rho_0 \mathbf{v} + \rho' \mathbf{v}) = 0 \end{aligned} \tag{3.22a}$$

According to Fig. 3.3, it gives $\frac{\partial \rho_0}{\partial t} << \frac{\partial \rho'}{\partial t}$, and $\nabla.(\rho' \mathbf{v}) << \nabla.(\rho_0 \mathbf{v})$
Thus, the continuity equation in Eq. (3.22a) becomes

$$\frac{\partial \rho'}{\partial t} + \nabla.(\rho_0 \mathbf{v}) = 0 \tag{3.22b}$$

where

ρ	is the absolute gas density given in Eq. (3.1b);
ρ'	is the perturbed gas density;
ρ_0	is the undisturbed gas density;
$\mathbf{v}$	is the fluctuating fluid velocity in the transmitting medium;
$\nabla.(\rho_0 \mathbf{v})$	is the divergence of the vector $\rho_0 \mathbf{v}$.

The Nabla operator ∇ is defined in

- Cartesian coordinate system (x,y,z):

$$\nabla = \frac{\partial}{\partial x}\mathrm{i} + \frac{\partial}{\partial y}\mathrm{j} + \frac{\partial}{\partial z}\mathrm{k} \tag{3.23a}$$

- Spherical coordinate system (r,θ,φ):

$$\nabla = \frac{\partial}{\partial r}\mathbf{e}_r + \frac{1}{r}\cdot\frac{\partial}{\partial \theta}\mathbf{e}_\theta + \frac{1}{r\sin\theta}\cdot\frac{\partial}{\partial \phi}\mathbf{e}_\phi \tag{3.23b}$$

Because $\rho' << \rho_0$, $\mathbf{v}$ is very small, and p_0 unchanged with locations, it results in

$$\rho' \mathbf{v} << \rho_0 \mathbf{v} \;\Rightarrow\; \frac{\partial(\rho \mathbf{v})}{\partial t} \approx \frac{\partial(\rho_0 \mathbf{v})}{\partial t}$$
$$p_0(x,y,z) = \text{const} \;\Rightarrow\; \nabla p = \nabla p'$$

The conservation of momentum equation of viscous gas flows is given by

$$\frac{\partial(\rho_0 \mathbf{v})}{\partial t} + \nabla.(\rho_0 \mathbf{v} \otimes \mathbf{v}) = -\nabla p' + \nabla.\boldsymbol{\tau}' \tag{3.24}$$

where

$\nabla p'$	is the noise pressure vector acting on the fluid;
$\nabla.\boldsymbol{\tau}'$	is the perturbed viscous shear tensor;
$\mathbf{v} \otimes \mathbf{v}$	is the tensor product of the gas velocity.

The relation of the timely change rates of propagating noise pressure and perturbed gas density in an isentropic process is resulted from Eq. (3.8).

$$\frac{\partial p'}{\partial t} = c_0^2 \frac{\partial \rho'}{\partial t} \tag{3.25}$$

Having combined Eqs. (3.22b and 3.25), it results in

$$\frac{\partial p'}{\partial t} + c_0^2 \nabla.(\rho_0 \mathbf{v}) = 0 \tag{3.26}$$

Timely differentiating Eq. (3.26), one achieves

$$\frac{\partial^2 p'}{\partial t^2} + c_0^2 \frac{\partial}{\partial t} \nabla.(\rho_0 \mathbf{v}) = \frac{\partial^2 p'}{\partial t^2} + c_0^2 \nabla. \left[\frac{\partial(\rho_0 \mathbf{v})}{\partial t} \right] = 0 \tag{3.27}$$

The convective term, the second term on the left-hand side of Eq. (3.24), is second order of the gas velocity **v** and therefore mostly negligible in linear aeroacoustics due to small oscillating air-particle velocities according to Eq. (3.13).

Thus, Eq. (3.24) becomes for inviscid gas flows

$$\frac{\partial(\rho_0 \mathbf{v})}{\partial t} \approx -\nabla p' \tag{3.28}$$

Substituting Eq. (3.28) into (3.27), one obtains the *acoustic propagation equation* linearized Euler's equation (LEE), in which the viscous shear in the gas flow is negligibly small, and the perturbation amplitudes are very low. Linear acoustics generally applies the LEE, derived from the Navier-Stokes equations at low Reynolds numbers of perfect gases in an isentropic process, to study the noise propagation in the transmitting fluid.

$$\frac{\partial^2 p'}{\partial t^2} - c_0^2 \nabla^2 p' \equiv \frac{\partial^2 p'}{\partial t^2} - c_0^2 \Delta p' = 0 \tag{3.29a}$$

where

- p' is the perturbed noise pressure;
- c_0 is the time-averaged or undisturbed sound speed, as given in Eqs (3.1a, 3.4, and 3.5);
- $\Delta p'$ is the Laplacian operator of the noise pressure p'.

The wave equation of a scalar velocity component v(x,y,z,t) is given by substituting Eq. (3.13) into Eq. (3.29a).

$$\nabla^2 \mathrm{v} - \frac{1}{c_0^2} \cdot \frac{\partial^2 \mathrm{v}}{\partial t^2} \equiv \Delta \mathrm{v} - \frac{1}{c_0^2} \cdot \frac{\partial^2 \mathrm{v}}{\partial t^2} = 0 \tag{3.29b}$$

The Laplacian operator $\Delta \varphi$ with $\varphi = p'$; v(x,y,z,t) is formulated in

- Cartesian coordinate system (x,y,z):

$$\Delta\varphi \equiv \nabla^2\varphi = \nabla.\nabla\varphi = \frac{\partial^2\varphi}{\partial x^2} + \frac{\partial^2\varphi}{\partial y^2} + \frac{\partial^2\varphi}{\partial z^2} \tag{3.30a}$$

- Spherical coordinate system (r,θ,ϕ) in the radial direction r:

$$\Delta\varphi \equiv \nabla^2\varphi = \nabla.\nabla\varphi = \frac{1}{r^2}\cdot\frac{\partial}{\partial r}\left(r^2\frac{\partial\varphi}{\partial r}\right) = \frac{\partial^2\varphi}{\partial r^2} + \frac{2}{r}\cdot\frac{\partial\varphi}{\partial r} \tag{3.30b}$$

Similarly, the *Lighthill's equation* (also Lighthill's analogy) is derived from the nonlinear Navier-Stokes equations of viscous gas flows at high Reynolds numbers and aeroacoustics. It has been considered as the bridge between aerodynamics and aeroacoustics for the acoustic field. The Lighthill's equation describes the acoustic propagation at large perturbation amplitudes of viscous gas flows with high Reynolds numbers in nonlinear aeroacoustics due to the nonlinear terms on the rhs [2].

$$\begin{cases} \dfrac{\partial^2\rho'}{\partial t^2} - \Delta p' = \nabla.\nabla.(\rho_0\mathbf{v}\otimes\mathbf{v} - \boldsymbol{\tau}') \Leftrightarrow \\ \dfrac{\partial^2 p'}{\partial t^2} - c_0^2\Delta\rho' = c_0^2\nabla.\nabla.(\rho_0\mathbf{v}\otimes\mathbf{v} - \boldsymbol{\tau}') + \dfrac{\partial^2}{\partial t^2}(p' - c_0^2\rho') \end{cases} \tag{3.31a}$$

within

$\rho_0\mathbf{v}\otimes\mathbf{v}$ is the convective tensor;
$\boldsymbol{\tau}'$ is the perturbed viscous shear tensor.

The non-isentropic source term of $\partial^2/\partial t^2(p' - c_0{}^2\rho')$ on the rhs of Eq. (3.31a) describes the nonlinear equivalent acoustic source in the Lighthill's equation. However, this term vanishes at small perturbation amplitudes in the isentropic noise propagation.

Having substituted $\Delta p' = c_0^2\Delta\rho' + \nabla.\nabla.\left((p' - c_0^2\rho')\mathbf{I}\right)$ into the first equation of (3.31a), the Lighthill's equation is given in the perturbed gas density.

$$\frac{\partial^2\rho'}{\partial t^2} - c_0^2\Delta\rho' = \nabla.\nabla.\mathbf{T} \tag{3.31b}$$

The Lighthill's turbulence tensor $\mathbf{T}$ is given by

$$\mathbf{T} \equiv \rho_0\mathbf{v}\otimes\mathbf{v} - \boldsymbol{\tau}' + (p' - c_0^2\rho')\mathbf{I} \tag{3.31c}$$

where

$\mathbf{I}$ is the unity tensor equivalent to the Kronecker delta tensor.

In the case of inviscid flows at low Reynolds numbers and small perturbation amplitudes in an isentropic process for linear acoustics, all rhs terms in Eqs. (3.31a and 3.31b) equal zero. Thus, the Lighthill's equation (3.31a) in the perturbed pressure becomes the LEE given in Eq. (3.29a).

The acoustic propagation equation LEE of the *average noise pressure* p' is written in the radial direction r.

$$\frac{\partial^2 p'}{\partial t^2} - \frac{c_0^2}{r^2}\frac{\partial}{\partial r}\left(r^2 \frac{\partial p'}{\partial r}\right) = 0 \tag{3.32}$$

Substituting a new unknown variable $m = p' \cdot r$ into Eq. (3.32), one obtains the acoustic propagation equation LEE in the radial direction after a few arithmetic calculations.

$$\frac{\partial^2 (m)}{\partial t^2} - c_0^2 \frac{\partial^2 (m)}{\partial r^2} = 0 \tag{3.33}$$

The noise pressure in Eq. (3.33) can be written for the forward and backward propagating waves in the complex formulation

$$\begin{aligned} p'(r,t) &= \frac{m}{r} = p'_{fw}(r,t) + p'_{bw}(r,t) \\ &= \frac{A_1}{r} e^{j\omega(t-r/c)} + \frac{A_2}{r} e^{j\omega(t+r/c)} \end{aligned} \tag{3.34}$$

where

A_i is the integrating coefficients;

ω is the oscillation angular frequency;

r is the noise transmitting distance.

The *wavenumber* k is defined by the ratio of the angular frequency of air particles ω to time-averaged or undisturbed sound speed c_0.

$$k \equiv \frac{\omega}{c_0} \approx \frac{\omega}{c} = \frac{2\pi}{\lambda} \tag{3.35}$$

where λ is the wavelength of acoustic propagation with $\lambda = cT$.

The noise pressure is generated by the spherical surface that vibrates in the radial direction r with a velocity $v = \omega\xi$ at frequency ω about its equilibrium radius a. The complex noise pressure in the forward propagating direction is resulted in [1].

$$p'(r,t) = \frac{1}{(1 + jka)} \cdot \frac{\rho_0 j\omega \mathbf{Q}}{4\pi r} e^{-jk(r-a)} \tag{3.36}$$

within the complex volume velocity of the noise source $\mathbf{Q}$ [m^3/s] is written in

$$\mathbf{Q} = Q^* e^{j\omega t} = (4\pi a^2 \mathrm{v}) e^{j\omega t} \tag{3.37}$$

The noise source strength is defined by $\rho_0 j\omega Q^*$ which is given in Eq. (3.38).

$$\rho_0 \dot{\mathbf{Q}} = \rho_0 \frac{d}{dt}(\mathrm{Q}^* e^{\mathrm{j}\omega t}) = \rho_0 \mathrm{j}\omega\, \mathrm{Q}^* e^{\mathrm{j}\omega t} = \rho_0 \mathrm{j}\omega\, \mathbf{Q} \tag{3.38}$$

The complex volume acceleration $d\mathbf{Q}/dt$ [m^3/s^2] causes the noise propagation at the velocity v by sound radiation in the transmitting medium.

For a point monopole source (i.e., $ka \ll 1$), the noise pressure is given by substituting Eq. (3.38) into Eq. (3.36).

$$\begin{aligned} p'(r,t) &= \frac{\rho_0 \mathrm{j}\omega\, \mathbf{Q}}{4\pi r} e^{-\mathrm{j}kr} = \frac{\rho_0 \dot{\mathbf{Q}}}{4\pi r} e^{-\mathrm{j}kr} \\ &= \frac{\rho_0 \mathrm{j}\omega\, \mathrm{Q}^*}{4\pi r} e^{\mathrm{j}(\omega t - kr)} \end{aligned} \tag{3.39}$$

The inverse square law of the noise intensity is resulted from calculating the root mean square of $p'(r,t)$.

$$I_{\mathrm{rms}} \propto p'^2_{\mathrm{rms}} = \frac{(\rho_0 ck)^2}{32\pi^2 r^2} Q^{*2} \tag{3.40}$$

This inverse square law indicates that the noise intensity decreases with the square transmitting radius r; i.e., the longer the distance from the noise source, the lower the noise level is generated according to Eq. (3.40).

3.3 Computation of Acoustic Propagations

3.3.1 Computational Methods

The noise computational simulation of automotive turbochargers is much more complex and difficult than computational fluid dynamics (CFD) because the airborne noise computation must be divided into many interdisciplinary simulations based on finite element method (FEM), finite volume method (FVM), and boundary element method (BEM), such as the noise generation simulations using CFD and computational nonlinear rotordynamics (CNR), in the near-field acoustic propagation simulations using computational aeroacoustics including LEE for linear aeroacoustics, acoustic perturbation equations (APE) using the Lighthill's equation for nonlinear aeroacoustics in the mid-field, and far-field simulations of noise radiation into the environment using the Ffowcs Williams and Hawkings' equation (FWHE) based on the integral method BEM, as shown in Fig. 3.4. The BEM solves the Kirchhoff-Helmholtz integral equation for 3-D acoustic field problems by summation of the noise pressures over the boundary surface of the noise source body.

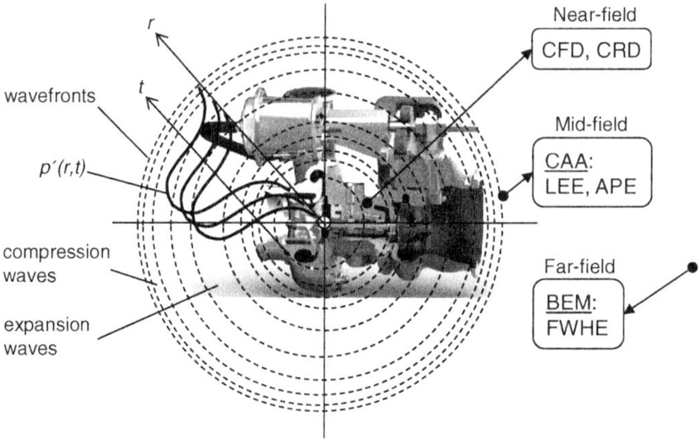

Fig. 3.4 Acoustic propagation computational methods

First, the generation of aerodynamic noise sources, such as flow vortexes and turbulent flows, is transiently computed by CFD with a direct numerical simulation (DNS) for which all flow initial and boundary conditions, and turbocharger geometries must be known in advance. CNR is applied to simulate the rotordynamic noise sources in the turbocharger (see Chap. 5). Second, using the LEE and APE the acoustic propagation through the components neighboring turbochargers are computed under mutual influences of the component walls on the noise levels. Third, the FWHE is used to compute the noise radiation neglecting the component wall effects at different rotor speeds in the far-field of the turbocharger (i.e. within a few meters from the vehicle).

The Lighthill's equations given in Eqs. (3.31a and 3.31b) with the noise source terms on the rhs are usually solved by the three-dimensional CAA program (3-D CAA) based on the FEM and BEM for complex systems [1–3]. As an example, such complex simulation models are applied in automotive industries where the noise sources induced from the turbocharger, as discussed in Sect. 3.2, are transmitted through the bearing center housing, compressor housing, charge-air intercoolers, air inlet system, exhaust-gas system (catalyzer, DPF, and muffler), and car frame to the cabin. Additionally, the strengths of aero- and rotordynamic noise sources Q_i on the rhs term of the Lighthill's equation are computed by 3-D CFD of flows in the compressor and turbine wheels [4] and the CNR of the turbocharger [5, 6], respectively.

The BEM and FEM are briefly outlined for linear aeroacoustics in the following subsections.

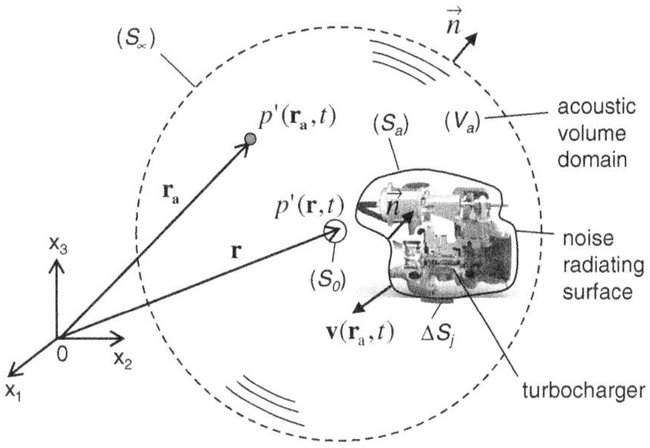

Fig. 3.5 BEM in an exterior acoustic field (V_a)

3.3.2 BEM for Aeroacoustics

The BEM based on Green's second theorem is used to transform the Helmholtz equation in the acoustic volume domain (V_a) into the Kirchhoff-Helmholtz integral equation defined on the closed surface (S_a). The noise propagating domain (V_a) is enclosed by the closed surfaces S_∞, S_0, and S_a, as shown in Fig. 3.5.

The noise pressure $p'(\mathbf{r},t)$ at any point $\mathbf{r}$ in the noise field (V_a) is solved by the Kirchhoff-Helmholtz integral equation at the given boundary conditions of $p'(\mathbf{r_a},t)$ and $\mathrm{v}(\mathbf{r_a},t)$ on the boundary surface (S_a).

The homogeneous noise pressure equation given in Eq. (3.29a) is written in the acoustic field volume domain.

$$\nabla^2 p' - \frac{1}{c_0^2}\frac{\partial^2 p'}{\partial t^2} = 0 \tag{3.41}$$

The timely second derivative of the harmonic noise pressure p' can be written in

$$p'(\mathbf{r},t) = \hat{p}'(\mathbf{r})\,\mathrm{e}^{\mathrm{j}\omega t} \Rightarrow \frac{\partial^2 p'(\mathbf{r},t)}{\partial t^2} = -\omega^2 p'(\mathbf{r},t) \tag{3.42}$$

The homogeneous Helmholtz equation of the noise pressure p' is resulted from Eqs. (3.35, 3.41, and 3.42).

$$\begin{aligned} &\nabla^2 p'(\mathbf{r},t) + k^2 p'(\mathbf{r},t) = 0;\ \forall \mathbf{r} \in V_a \\ &k \equiv \omega / c_0 \approx 2\pi/\lambda \end{aligned} \tag{3.43}$$

In the case of a noise source $\mathrm{Q}(\mathbf{r},t)$ in the acoustic field V_a, the inhomogeneous Helmholtz equation is formulated in

$$\nabla^2 p'(\mathbf{r},t) + k^2 p'(\mathbf{r},t) = \mathrm{Q}(\mathbf{r},t);\ \forall \mathbf{r}, \mathrm{Q} \in V_a \tag{3.44a}$$

where the noise source in the rhs is defined by [1]

$$Q(\mathbf{r},t) = \nabla.\mathbf{f}(\mathbf{r},t) - \mathrm{j}\rho_0\omega\, q(\mathbf{r},t) \tag{3.44b}$$

The rhs term in Eq. (3.44a) describes the force and noise source intensity acting on the acoustic field volume domain, as described in the Lighthill's equation (3.31b and 3.31c) according to [1].

Generally, there are three kinds of boundary conditions at the enclosed surface (*S*) for the Helmholtz equation:

- Dirichlet boundary condition at the rigid surface

$$\left.\frac{\partial p'}{\partial n}\right|_{(S)} = 0 \Rightarrow p'_{(S)}(\mathbf{r_a},t) = \text{constant} \tag{3.44c}$$

- Neumann boundary condition at the vibrating surface

$$\left.\frac{\partial p'}{\partial n}\right|_{(S)} = -\mathrm{j}\rho_0\omega\ \mathrm{v}_{(S)}(\mathbf{r_a},t) \tag{3.44d}$$

- Robin boundary condition (also impedance BC for Sturm-Liouville problems)

$$\left.\frac{\partial p'}{\partial n}\right|_{(S)} = -\mathrm{j}\rho_0\omega\,\frac{p'_{(S)}(\mathbf{r_a},t)}{Z};\ Z \equiv \rho_0 c \approx \frac{\rho_0\omega}{k} \tag{3.44e}$$

where Z is the noise impedance of the acoustic field, as given in Eq. (3.13).

Green's second theorem transforms the volume domain V into the boundary S.

$$\oint_S \left(\varphi\frac{\partial\psi}{\partial n} - \psi\frac{\partial\varphi}{\partial n}\right)dS = \int_V (\varphi\nabla^2\psi - \psi\nabla^2\varphi)dV \tag{3.45}$$

where φ and ψ are two scalar functions defined in the volume domain V.

To apply the BEM to linear aeroacoustics, the noise pressure $p'(\mathbf{r_a},t)$ at the location $\mathbf{r_a}$ in the acoustic field V_a is substituted for φ; the free-field Green's function $G(\mathbf{r},\mathbf{r_a})$ in the acoustic field V_a with the wavenumber k is used for ψ [1, 7].

$$\begin{cases} \phi \equiv p'(\mathbf{r_a},t); \\ \psi \equiv G(\mathbf{r},\mathbf{r_a}) = \frac{e^{-\mathrm{j}k|\mathbf{r}-\mathbf{r_a}|}}{4\pi|\mathbf{r}-\mathbf{r_a}|};\quad k \equiv \frac{\omega}{c_0} \approx \frac{2\pi}{\lambda} \end{cases} \tag{3.46}$$

Having applied the Green's second theorem to the closed boundary surfaces of S_∞, S_Q, and S_a, one obtains the Kirchhoff-Helmholtz integral equation by setting the rhs term in Eq. (3.45) to equal zero because both $p'(\mathrm{r_a},t)$ and $G(\mathbf{r},\mathbf{r}_\mathrm{a})$ satisfy the homogeneous Helmholtz equation (3.43).

$$p'(\mathbf{r},t) = \frac{1}{c(\mathbf{r})} \oint_{S_a} \left[p'(\mathbf{r_a},t).\frac{\partial G(\mathbf{r},\mathbf{r_a})}{\partial n} - G(\mathbf{r},\mathbf{r_a}).\frac{\partial p'(\mathbf{r_a},t)}{\partial n} \right] dS_a \tag{3.47a}$$

where the domain coefficient $c(\mathbf{r})$ is resulted in [1]

$$c(\mathbf{r}) = \begin{cases} -1 \text{ if } \forall \mathbf{r} \in V_a \\ -\frac{1}{2} \text{ if } \forall \mathbf{r} \in S_a \\ 0 \text{ if } \forall \mathbf{r} \notin V_a \end{cases} \tag{3.47b}$$

The vibrating velocity of the solid surface is given by the Neumann boundary condition given in Eq. (3.44d).

$$\begin{aligned} &\left.\frac{\partial p'(\mathbf{r_a},t)}{\partial n}\right|_{(S_a)} = -\mathrm{j}\rho_0\omega\ \mathrm{v}_{(S_a)} \approx -\mathrm{j}\rho_0 ck\ \mathrm{v}_{(S_a)} \\ &\Rightarrow \mathrm{v}(\mathbf{r_a},t)|_{(S_a)} = \frac{\mathrm{j}}{\rho_0 ck}.\left.\frac{\partial p'(\mathbf{r_a},t)}{\partial n}\right|_{(S_a)} \end{aligned} \tag{3.48}$$

Integrating the rhs term of Eq. (3.47a) at the noise pressure $p'(\mathbf{r_a},t)$ and vibrating velocity $\mathrm{v}(\mathbf{r_a},t)$ on the boundary surface S_a, it gives the noise pressure at any point $\mathbf{r}$ in the noise field V_a. In the case of $\mathbf{r}$ inside V_a at $c(\mathbf{r}) = -1$, the induced noise pressure results in

$$p'(\mathbf{r},t) = \oint_{S_a} \left[G(\mathbf{r},\mathbf{r_a}).\frac{\partial p'(\mathbf{r_a},t)}{\partial n} - p'(\mathbf{r_a},t).\frac{\partial G(\mathbf{r},\mathbf{r_a})}{\partial n} \right] dS_a \tag{3.49}$$

The noise pressure p' is written in the Helmholtz operators by substituting the additional incident acoustic field p'_{in} into Eq. (3.49).

$$p'(\mathbf{r},t) = \{Lv\}_{S_a} - \{Mp'\}_{S_a} + p'_{\mathrm{in}}(\mathbf{r},t);\ \forall \mathbf{r} \in V_a \tag{3.50a}$$

In which the incident noise pressure p'_{in} is resulted from integrating the source term $\mathrm{Q}(\mathbf{r},t)$ over the acoustic field volume V_S.

$$p'_{in}(\mathbf{r},t) = \oint_{V_S} \mathrm{Q}(\mathbf{r}_s,t).G(\mathbf{r},\mathbf{r_s})\ dV_S \tag{3.50b}$$

The Helmholtz integral operators on the boundary surface S_a are defined by

$$\begin{cases} \{Lv\}_{S_a} \equiv \oint_{S_a} \left[G(\mathbf{r},\mathbf{r_a}).\frac{\partial p'(\mathbf{r_a},t)}{\partial n} \right] dS_a; \\ \{Mp'\}_{S_a} \equiv \oint_{S_a} \left[p'(\mathbf{r_a},t).\frac{\partial G(\mathbf{r},\mathbf{r_a})}{\partial n} \right] dS_a \end{cases} \tag{3.51}$$

The integrated boundary surface S_a is divided into N finite boundary elements $\Delta S_{j=1,\ldots,N}$, as shown in Fig. 3.5. Thus, the noise pressure of Eq. (3.50a) inside the acoustic field V_a is resulted by integrating the rhs of Eq. (3.50a) over the finite boundary surfaces $\Delta S_{j=1,\ldots,N}$.

$$p'(\mathbf{r},t) = \sum_{j=1}^{N} \{L_k \mathrm{e}\}_{\Delta S_j} \mathrm{v_j}(\mathbf{r_a},t) - \sum_{j=1}^{N} \{M_k \mathrm{e}\}_{\Delta S_j} p'_j(\mathbf{r_a},t) + p'_{\mathrm{in}}(\mathbf{r},t); \quad \forall \mathbf{r} \in V_a \tag{3.52}$$

where e is the unit vector on the finite boundary element ΔS_j.

The Helmholtz integral operators in Eq. (3.52) are given by

$$\begin{cases} \{L_k \mathrm{e}\}_{\Delta S_j \subset S_a} \equiv \int_{\Delta S_j} G_k(\mathbf{r},\mathbf{r_a}) dS_a; \\ \{M_k \mathrm{e}\}_{\Delta S_j \subset S_a} \equiv \int_{\Delta S_j} \frac{\partial G_k(\mathbf{r},\mathbf{r_a})}{\partial n} dS_a \end{cases} \tag{3.53}$$

Due to the solution singularity for certain values of the wavenumber k on the surface domain, other integrating methods, such as Schenck method based on combined Helmholtz Integral equation formulation (CHIEF), improved indirect, and direct formulations are used to compute the noise pressure in the acoustic field [1, 2, 7]. The commercial 3-D Acoustics Code of LMS Virtual Lab [8] based on the boundary element acoustics (BEA) is mostly used to compute three-dimensional acoustic field problems in automotive applications.

3.3.3 FEM for Aeroacoustics

Instead of solving the Helmholtz equation by integrating the noise pressure over the radiating surface (S_a) by means of BEM, the FEM directly solves this equation in a three-dimensional acoustic filed (V) surrounding the radiating surface of the turbocharger, as shown in Fig. 3.6.

Multiplying the Helmholtz equation by a scalar function ϕ differentiable in the control volume V and integrating both sides of the equation over dV, one achieves the weak Galerkin formulation of the Helmholtz equation:

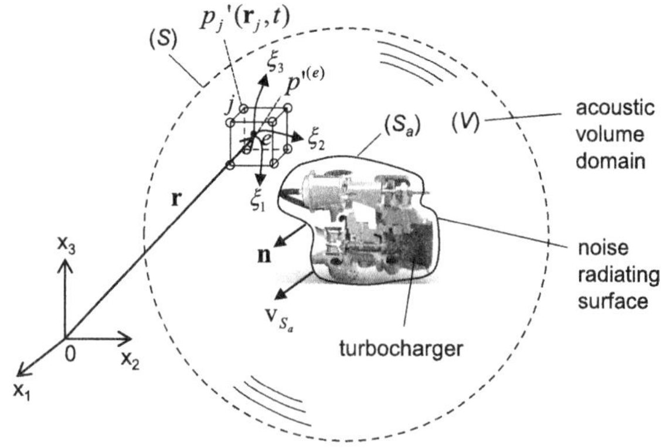

Fig. 3.6 FEM in an acoustic field (V)

$$\int_V \phi\, \nabla.\nabla p'\, dV + \int_V \phi\, k^2 p'\, dV = 0 \tag{3.54}$$

Having used the product chain rule of the divergence

$$\nabla.(\phi.\nabla p') = \phi.\nabla.\nabla p' + \nabla\phi.\nabla p',$$

Equation (3.54) is rewritten in

$$\int_V \nabla.(\phi\nabla p')dV - \int_V \nabla\phi.\nabla p'\, dV + \int_V \phi\, k^2 p'\, dV = 0 \tag{3.55}$$

Applying Gauss divergence theorem to the control volume V, it is given by

$$\begin{aligned} \int_V \nabla.(\phi\nabla p')dV &= \oint_S (\phi\nabla p').\mathbf{n}\, dS \\ &= \oint_S \phi \frac{\partial p'}{\partial n} dS = -\mathrm{j}\rho_0\omega\, \mathrm{v}_{S_a} \int_{S_a} \phi\, dS \end{aligned} \tag{3.56}$$

Substituting Eq. (3.56) into Eq. (3.55), one obtains

$$\int_V (\nabla\phi.\nabla p' - k^2\phi\, p')dV = -\mathrm{j}\rho_0\omega\, \mathrm{v}_{S_a} \int_{S_a} \phi\, dS \tag{3.57}$$

within the fluid velocity v_{Sa} on the radiating surface satisfies the Neumann boundary condition given in Eq. (3.44d).

$$\mathrm{v}_{S_a} = \frac{\mathrm{j}}{\rho_0\omega}.\frac{\partial p'}{\partial n}\bigg|_{(S_a)} \tag{3.58}$$

The Rayleigh-Ritz method approaches the noise pressure $p'^{(e)}$ in the finite element e = 1,…,N_e in the local coordinate system (ξ_1, ξ_2, ξ_3) by a finite series expansion according to [1].

$$\begin{aligned} p' \equiv p'^{(e)} &\approx \{\mathbf{S}(\xi_1, \xi_2, \xi_3)\}.\{\boldsymbol{\alpha}\} = \left(\{\mathbf{S}(\xi_1, \xi_2, \xi_3)\}.[\mathbf{D_e}]^{-1}\right).\{\mathbf{p}'\} \\ &= \{\boldsymbol{\psi}(\xi_1, \xi_2, \xi_3)\}.\{\mathbf{p}'\} \approx \sum_{j=1}^{N} \psi_j(\xi_1, \xi_2, \xi_3) p'_j \end{aligned} \tag{3.59}$$

where

- $\mathbf{S}$ is the shape function ($1 \times N$) matrix;
- $\boldsymbol{\alpha}$ is the coefficient ($N \times 1$) vector;
- $\mathbf{D_e}$ is the distribution function ($N \times N$) matrix;
- $\boldsymbol{\psi}$ is the weighting function ($1 \times N$) matrix;
- $\mathbf{p}'$ is the noise pressure ($N \times 1$) vector;

N_e is the number of discretized finite elements in the control volume V;
N is the nodal points of the discretized finite element;
ψ_j is the pressure weighting function at the nodal point j that denotes the weighting rate of the noise pressure per unit volume;
p'_j is the noise pressure at the nodal point j.

Inserting Eq. (3.59) into Eq. (3.57), it gives at choosing $\phi \equiv \psi_i$.

$$\sum_{j=1}^{N}\left[\int_V (\nabla\psi_i.\nabla\psi_j - k^2\psi_i\psi_j)\,dV\right].p'_j = -j\rho_0\omega\, v_{S_a}\int_{S_a}\psi_i dS;\ \forall i = 1,\ldots,N \tag{3.60}$$

Thus, Eq. (3.60) is formulated in the matrix equation of the noise pressure.

$$[\mathbf{A}].\{\mathbf{p}'\} = \{\mathbf{q}\} \tag{3.61a}$$

where
$[\mathbf{A}] = (N \times N)$ matrix of the components

$$a_{ij} \equiv \int_V (\nabla\psi_i.\nabla\psi_j - k^2\psi_i\psi_j)\,dV;\ \forall i = 1, 2,\ldots,N;\ \forall j = 1, 2,\ldots,N \tag{3.61b}$$

$\{\mathbf{p}'\} = (N \text{ x } 1)$ vector of the noise pressures

$$\{p'\} = \left[p'_1\, p'_i \ldots p'_N\right]^T;\ \forall i = 1, 2,\ldots,N \tag{3.61c}$$

$\{\mathbf{q}\} = (N \times 1)$ vector of $[q_1\ q_2\ \ldots\ q_N]^{\mathrm{T}}$ of the components

$$q_i = -j\rho_0\omega\, v_{S_a}\int_{S_a}\psi_i dS;\ \forall i = 1, 2,\ldots,N \tag{3.61d}$$

The weighting functions ψ of the considered finite element depend on the discretizing scheme [1, 7]:

- For a local orthogonal curvilinear coordinate (ξ_1, ξ_2, ξ_3), the weighting functions ψ_j (ξ_1, ξ_2, ξ_3) based on a quadratic element with $N = 4$ nodal points are given

$$\{\boldsymbol{\psi}_{4NP}(\xi_1,\xi_2,\xi_3)\} = [\psi_1 \quad \psi_2 \quad \psi_3 \quad \psi_4] \tag{3.62}$$

- For a local orthogonal curvilinear coordinate (ξ_1, ξ_2, ξ_3), the weighting functions ψ_j (ξ_1, ξ_2, ξ_3) based on a hexahedral element with $N = 8$ nodal points are given

$$\{\psi_{8NP}(\xi_1, \xi_2, \xi_3)\} = [\psi_1 \quad \psi_2 \quad \psi_3 \quad \psi_4 \quad \psi_5 \quad \psi_6 \quad \psi_7 \quad \psi_8] \tag{3.63}$$

To transform the integral Eqs. (3.60, 3.61b and 3.61d) from the orthogonal curvilinear coordinate (ξ_1, ξ_2, ξ_3) of the finite element mesh into the Cartesian coordinate (x_1, x_2, x_3), the Jacobian determinant **J** is used to compute the gradients of ψ and dV in the curvilinear coordinate (see Appendix B). The commercial 3-D Acoustics Code ACTRAN Aero and Vibroacoustics of the MSC Software [10] based on the FEM is mostly used to compute three-dimensional acoustic field problems in automotive applications.

The following computing steps must be carried out to find the noise pressure solution of Eq. (3.61a) in the acoustic field V:

- Compute the terms a_{ij} in Eq. (3.61b) and the rhs term q_i in Eq. (3.61d);
- Compute the noise pressures p'_j at the nodal points by solving the matrix equation (3.61a) at the boundary conditions given in Eqs. (3.44c–e);
- Compute the mean noise pressure in the discretized finite elements according to Eq. (3.59).

3.4 Frequency Modulations of Nonlinear Aeroacoustics

Nonlinear aeroacoustics deals with large noise perturbation amplitudes in which the acoustic characteristics are no longer linear. Therefore, synchronous and asynchronous responses occur in a nonlinear aeroacoustic system when first, the aero- and rotordynamic noise responses are superimposed to generate sidebands and; second, the airborne noise response is additionally generated by superimposing many acoustic propagation waves at different frequencies during propagation in the far-field at high Reynolds numbers.

The sidebands occurring at the sub- and supersynchronous frequencies are generated by the propagated noise waves at different frequencies. Note that the more visibly the sidebands appear in the sound spectra plot (Waterfall plot), the more nonlinearly the aeroacoustic system responds. The response amplitude of the nonlinear aeroacoustic system is resulted from all component amplitudes of the propagating noise waves and sidebands combined together. As a result, the noise response amplitude is intensified; the aeroacoustic behavior becomes much more nonlinear at large perturbation amplitudes in turbulent flows at high Reynolds numbers.

3.4.1 Responses of Nonlinear Aeroacoustic Systems

Having considered a nonlinear aeroacoustic system, the system response $y(x)$ is resulted from the input signal $x(t)$ with large amplitudes and the system transfer impedance $Z(\varepsilon)$, as shown in Fig. 3.7.

The nonlinear response $y(x)$ of the system is written in second order of the input function $x(t)$.

$$\begin{aligned} y(x) &= f(x, \varepsilon) = Z\left(x + \varepsilon.\mathrm{sgn}(x)x^2 + \ldots\right) \\ &= Z(x + \varepsilon^* x^2 + \ldots) \end{aligned} \tag{3.64}$$

where

x	is the input signal with large amplitudes;
Z	is the complex transfer impedance of the system;
Zx	is the linear term of the response;
ε	is the nonlinear system factor ($\varepsilon \ll 1$);
$Z\varepsilon^* x^2$	is the nonlinear term of the response;
$\varepsilon^* = \varepsilon.\mathrm{sgn}(x)$	within $\mathrm{sgn}(x)$ is the signum function of x, which is defined by

$$\mathrm{sgn}(x) = \begin{cases} -1 \text{ if } x < 0 \\ 0 \text{ if } x = 0 \\ +1 \text{ if } x > 0 \end{cases}$$

Therefore,

$$\varepsilon^* \equiv \varepsilon \mathrm{sgn}(x) = \begin{cases} -\varepsilon \text{ if } x < 0 \\ 0 \text{ if } x = 0 \\ +\varepsilon \text{ if } x > 0 \end{cases}$$

Thus, the nonlinear vibration response becomes

$$y(x) = Zx + Z\varepsilon \mathrm{sgn}(x)x^2 + O(x^3) \tag{3.65}$$

The nonlinear vibration response $y(x)$ is displayed in Fig. 3.8 and compared to the linear response. In the linear system, the response of Zx is proportional to the input function x. On the contrary, the nonlinear response is different from the linear term of Zx; it increases or decreases more quickly with the input amplitude $x(t)$ because the nonlinear term is proportional to x^2. As a result, the nonlinear system can only be linearized at the small amplitudes because the nonlinear term is quite small.

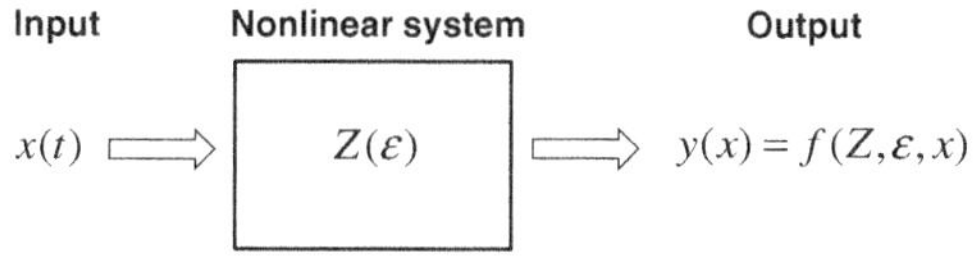

Fig. 3.7 Response of a nonlinear aeroacoustic system

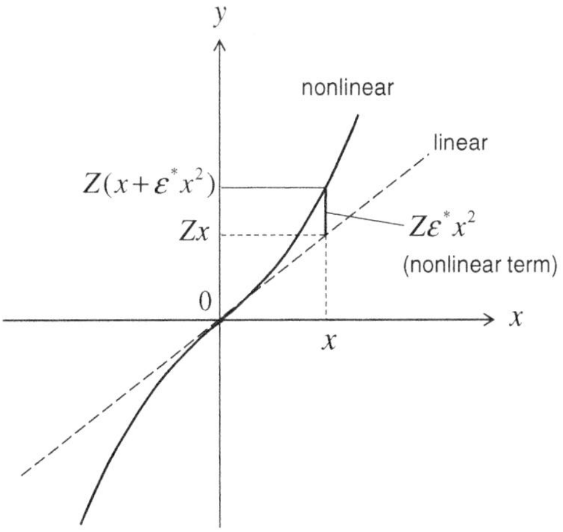

Fig. 3.8 Comparison of the linear and nonlinear responses

In case of the excessively large unbalance whistle amplitude, the system behavior is strongly nonlinear. Thus, the response amplitude $y(x)$ increases faster than the amplitude of the linear system since the nonlinear term increases with the input amplitude squared that is no longer negligible, as shown in Fig. 3.8.

3.4.2 Modulated Noise Sideband Frequencies

Having supposed that the input $x(t)$ is a harmonic acoustic propagation wave with the amplitude X and single frequency ω_1.

$$x = X \cos \omega_1 t \tag{3.66}$$

Its response is resulted from Eq. (3.64) in second order of $x(t)$.

$$\begin{aligned} y &= ZX \cos \omega_1 t + Z\varepsilon^* X^2 \cos^2 \omega_1 t \\ &\equiv y_{L-\text{term}} + y_{NL-\text{term}} \end{aligned} \tag{3.67}$$

The noise response consists of two terms, the first term $y_{L\text{-term}}$ is linear with $x(t)$ and the latter $y_{NL\text{-term}}$ is the nonlinear term of x^2. The nonlinear term can be written in another trigonometric formulation.

$$y_{L-\text{term}} \equiv ZX \cos \omega_1 t; \tag{3.68}$$

$$\begin{aligned} y_{NL-\text{term}} &\equiv Z\varepsilon^* x^2 = Z\varepsilon^* X^2 \cos^2 \omega_1 t \\ &= \frac{1}{2} Z\varepsilon^* X^2 (1 + \cos 2\omega_1 t) \end{aligned} \tag{3.69}$$

By substitution of Eqs. (3.67, 3.68, and 3.69), one gets the overall noise response.

$$y = ZX\cos\omega_1 t + \frac{1}{2}Z\varepsilon^* X^2 + \frac{1}{2}Z\varepsilon^* X^2\cos 2\omega_1 t \quad (3.70)$$

Compared to the linear system, the nonlinear response has two new additional terms: the rectification term of X^2 and the second-order harmonic term of $2\omega_1$.

Now, we consider the input function $x(t)$ of a periodic noise containing two frequency components of ω_1 and ω_2 and corresponding amplitudes X_1 and X_2.

$$x = X_1\cos\omega_1 t + X_2\cos\omega_2 t;\ \omega_1 > \omega_2 \quad (3.71)$$

To simplify, we consider only the nonlinear term of the noise response in the following calculation:

$$\begin{aligned} y_{NL-\text{term}} &\equiv Z\varepsilon^* x^2 = Z\varepsilon^*(X_1\cos\omega_1 t + X_2\cos\omega_2 t)^2 \\ &= Z\varepsilon^*(X_1^2\cos^2\omega_1 t + X_2^2\cos^2\omega_2 t + 2X_1X_2\cos\omega_1 t.\cos\omega_2 t) \\ &\equiv y_1 + y_2 + y_3 \end{aligned} \quad (3.72)$$

Similarly, the first two terms y_1 and y_2 provide two additional second-order harmonics of the frequencies $2\omega_1$ and $2\omega_2$, respectively. Furthermore, one obtains the third term y_3 that generates sidebands of ω_1 and ω_2 after a few trigonometric calculating steps.

$$\begin{aligned} y_3 &= Z\varepsilon^* X_1X_2(2\cos\omega_1 t.\cos\omega_2 t) \\ &= Z\varepsilon^* X_1X_2[\cos(\omega_1 t - \omega_2 t) + \cos(\omega_1 t + \omega_2 t)] \\ &= Z\varepsilon^* X_1X_2[\cos(\omega_1 - \omega_2)t + \cos(\omega_1 + \omega_2)t] \\ &\equiv Z\varepsilon^* X_1X_2(\cos\omega_{\text{LSB}}t + \cos\omega_{\text{USB}}t) \end{aligned} \quad (3.73)$$

with

$$\omega_{\text{USB}} = \omega_1 + \omega_2;\ \omega_{\text{LSB}} = \omega_1 - \omega_2 \quad (3.74)$$

where

ω_{USB} is the upper noise sideband frequency (rad/s);
ω_{LSB} is the lower noise sideband frequency (rad/s).

Thus,

$$\omega_1 = \frac{1}{2}(\omega_{\text{USB}} + \omega_{\text{LSB}}) \equiv \omega_c = 2\pi f_c; \quad (3.75)$$

with f_c is defined by the carrier frequency (1/s);

$$\omega_2 = \frac{1}{2}(\omega_{\text{USB}} - \omega_{\text{LSB}}) \equiv \omega_m = 2\pi f_m; \quad (3.76)$$

with f_m is defined by the modulation frequency (1/s).

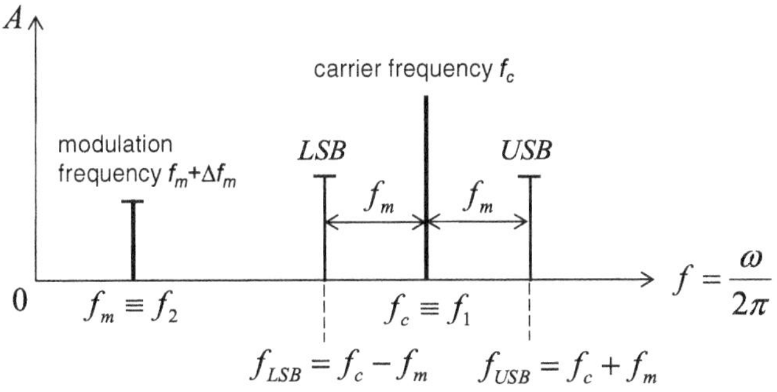

Fig. 3.9 Double sidebands of noise frequency modulations

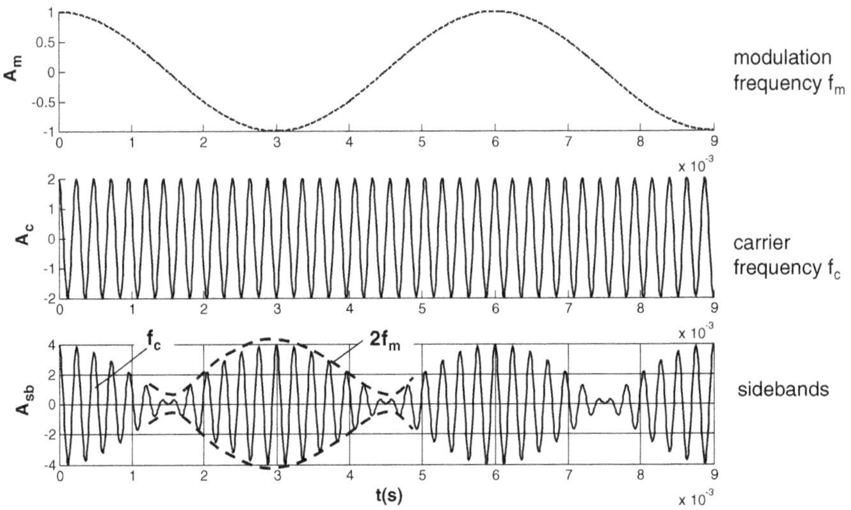

Fig. 3.10 Double sideband (*DSB*) of the amplitude modulation

Figure 3.9 shows the noise sideband frequencies of the modulation of the acoustic propagation waves. Due to the acoustic nonlinearity, the lower and upper sidebands are induced by the amplitude and frequency modulations of f_c and f_m. In fact, the noise sideband frequency varies in a bandwidth of frequency Δf_m, e.g., in case of oil whirl. Therefore, the noise sideband frequencies f_{LSB} and f_{USB} scatter in the corresponding frequency bandwidth Δf_m.

In the radio engineering, the carrier frequency f_c is much higher than the modulation frequency f_m ($f_c \gg f_m$). To broadcast the modulation signal in a long distance, the carrier signal modulates it in the double sideband (DSB) containing the lower (LSB) and upper sidebands (USB) with the ultra-high frequencies f_{LSB} and f_{USB} (see Figs. 3.9 and 3.10). The noise sideband frequencies are modulated by the carrier and modulation frequencies in the wideband of frequency.

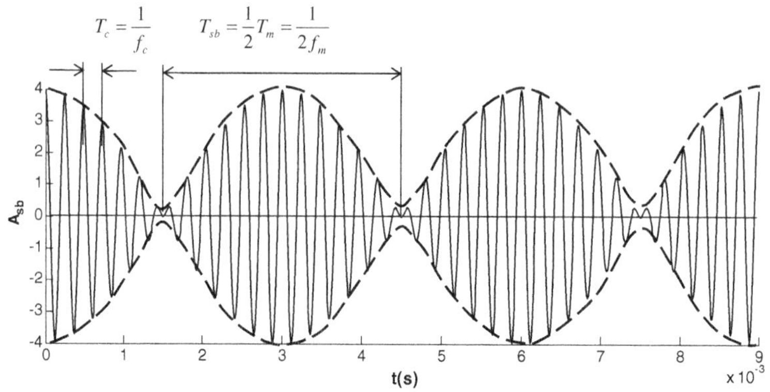

Fig. 3.11 Frequencies and periods of the sideband signal

The modulation frequency f_{m} is normally the human voice frequency spreading from 300 Hz to 3 kHz; therefore, the sidebands are generated with the maximum frequency width of the DSB of 2 f_{m} about 6 kHz. The modulation signal is recovered at receiving by using demodulation or detection of the received sideband signals where the signal detector and low-pass filter are used. The entire process of sending and receiving signal is carried out by the MODEM method that consists of Modulation and DEModulation of the signals, generally applied in the wideband technique.

The sideband signal generated by the carrier and modulating signals is displayed in Fig. 3.9, in which one finds the carrier and modulation frequencies again in the sideband signal. The envelop signal of the sideband given in Eq. (3.73) has a frequency of 2 f_{m} since its period equals one-half of the period of the modulation signal, as shown in Fig. 3.11.

The lower and upper noise sideband frequencies of f_{LSB} and f_{USB} are calculated from the carrier f_{c} and modulation frequencies f_{m} with $f_{\mathrm{m}} \ll f_{\mathrm{c}}$; hence, $f_1 \gg f_2$:

- Lower noise sideband frequency (LSB):

$$f_{\mathrm{LSB}} = f_c - f_m \equiv f_1 - f_2 \tag{3.77}$$

- Upper noise sideband frequency (USB):

$$f_{\mathrm{USB}} = f_c + f_m \equiv f_1 + f_2 \tag{3.78}$$

To sum up, the nonlinear acoustic system induces some new additional terms due to the amplitude and frequency modulations as follows:

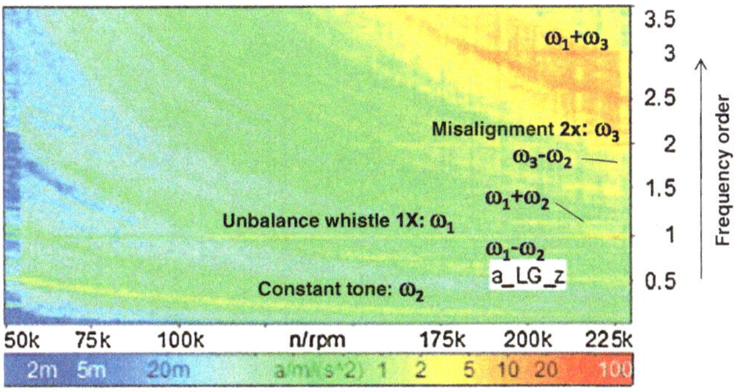

Fig. 3.12 Noise sideband frequencies in Waterfall plot (courtesy BMTS)

- The rectification term $\frac{1}{2}Z\varepsilon^* X^2$ in Eq. (3.70);
- Second-order harmonic frequencies of $2\omega_1$ and $2\omega_2$ with the amplitudes X_1^2 and X_2^2 in Eqs. (3.70) and (3.72);
- Noise sideband frequencies modulated by the excitation frequencies of ω_1 and ω_2 have the amplitudes that are proportional to the term of $Z\varepsilon^* X_1 X_2$. According to Eqs. (3.73 and 3.74), the noise sideband frequencies are the sum or difference of the frequency components; hence, the first-order noise sideband frequencies result in $3\omega_2 - 2\omega_1$, $2\omega_2 - \omega_1$, $2\omega_1 - \omega_2$, $3\omega_1 - 2\omega_2$, $4\omega_1 - 3\omega_2$, etc., (see Eqs. 3.80, 3.81), and Fig. 3.15.

In the case of the unbalanced and misaligned rotor, the sidebands induced by the frequency modulations of misalignment, unbalance whistle, and constant tone (inner oil whirl) are indicated in the noise spectra Waterfall plot, as displayed in Fig. 3.12.

Beat phenomenon, a special case of the modulation, is caused by the modulation of two harmonic vibrations with the frequency components of ω_1 and ω_2 that are close to each other. If their amplitudes are equal, only the amplitude modulation occurs. Otherwise, both amplitude and frequency modulations take place at the same time. The input signal is the sum of two sinus functions with the amplitudes $a_1 = a_2 \equiv a$.

$$
\begin{aligned}
x &= a_1 \sin \omega_1 t + a_2 \sin \omega_2 t \\
&= a(\sin \omega_1 t + \sin \omega_2 t) \\
&= 2a \sin\left[\frac{1}{2}(\omega_1 + \omega_2)\right] t \cdot \cos\left[\frac{1}{2}(\omega_1 - \omega_2)\right] t \\
&\equiv 2a \sin \omega_c t \cdot \cos \omega_m t
\end{aligned}
\tag{3.79}
$$

where

$\omega_c = \frac{1}{2}(\omega_1 + \omega_2) = 2\pi f_c;\ f_c$ is the carrier frequency.
$\omega_m = \frac{1}{2}(\omega_1 - \omega_2) = 2\pi f_m;\ f_m$ is the modulation frequency.

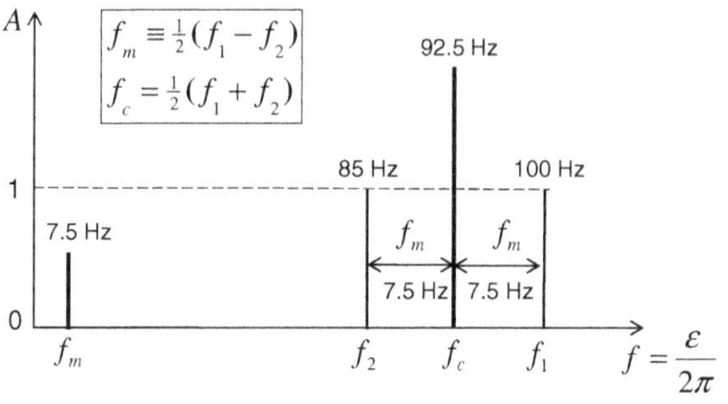

Fig. 3.13 Beat frequency $2\,f_m$ of the frequencies f_1 and f_2

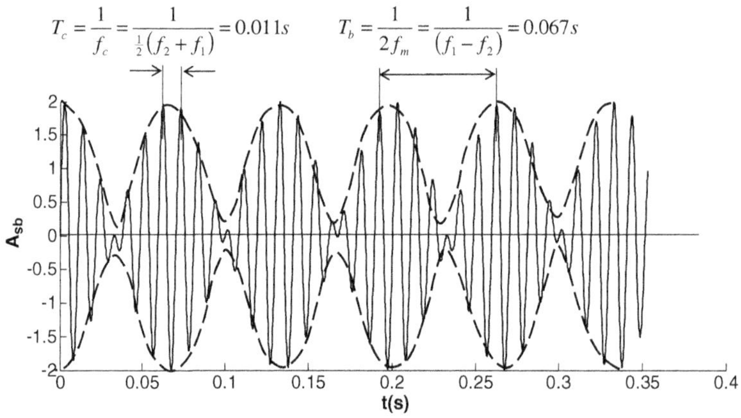

Fig. 3.14 Beat response in the time domain

To demonstrate the beat frequency, one chooses two frequency components of $f_1 = 100$ Hz and $f_2 = 85$ Hz for the beat modulation that is shown in the frequency domain (Fig. 3.13). The carrier frequency f_c is the average frequencies of f_1 and f_2 and equals 92.5 Hz, and the modulation frequency f_m equals 7.5 Hz. The beat frequency of $2\,f_m$ (15 Hz) is defined by the frequency difference of f_1 and f_2.

Figure 3.14 indicates that the closer the frequency components, the smaller the beat frequency; obviously, the larger the period of the beat signal. The beat period is inversely related to the beat frequency $2\,f_m$; hence, the beat period is about 0.067 s at the beat frequency of 15 Hz. The carrier frequency has a period of 0.011 s at 92.5 Hz in the time domain.

The generated first-order frequencies of the LSB and USB sidebands are resulted from the sum or difference of the frequency components of f_1 and f_2, which are located on the left- and right-hand sides of the carrier frequency f_c, as plotted in Fig. 3.15.

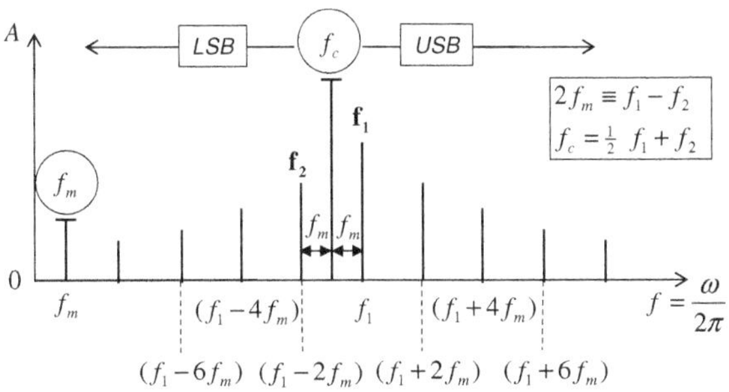

Fig. 3.15 LSB and USB noise sideband frequencies

Table 3.1 Noise sideband frequency modulations

Lower sideband (LSB)	f_c	Upper sideband (USB)	f Order
$2f_2 - f_1$, f_2	$(f_1 + f_2)/2$	f_1, $2f_1 - f_2$	1X
$3f_2 - f_1$, $2f_2$	$(f_1 + f_2)$	$2f_1$, $3f_1 - f_2$	2X
$4f_2 - f_1$, $3f_2$, $2f_2 + f_1$	$3(f_1 + f_2)/2$	$2f_1 + f_2$, $3f_1$, $4f_1 - f_2$	3X

$$f_{LSB} = \begin{cases} f_1 - 2f_m = f_c - f_m = f_2 \\ f_1 - 4f_m = f_c - 3f_m = 2f_2 - f_1 \\ f_1 - 6f_m = f_c - 5f_m = 3f_2 - 2f_1 \\ \ldots \end{cases} \tag{3.80}$$

and

$$f_{USB} = \begin{cases} f_1 + 0f_m = f_c + f_m = f_1 \\ f_1 + 2f_m = f_c + 3f_m = 2f_1 - f_2 \\ f_1 + 4f_m = f_c + 5f_m = 3f_1 - 2f_2 \\ f_1 + 6f_m = f_c + 7f_m = 4f_1 - 3f_2 \\ \ldots \end{cases} \tag{3.81}$$

However, the nonlinear terms of the noise response are not only proportional to the input amplitude squared x^2 but also to the high-order input amplitude, such as x^3 or higher orders compared to Eq. (3.64). According to Ehrich [9], the truncation of the modulation frequencies of f_1 and f_2 could produce two new second harmonic frequencies of $(f_1 - f_2)$ and $(f_1 + f_2)$ that in turn generate the third harmonics of $(2f_2 + f_1)$ and $(2f_1 + f_2)$ in a strongly nonlinear acoustic system, as demonstrated in Table 3.1.

To generate the LSB frequencies, one superimposes the excitation frequency f_2 on $(2f_2 - f_1)$ to obtain the second-order frequency $(3f_2 - f_1)$ following the thin arrow, and superimposing f_2 again on the second-order frequency to generate the third-order frequency $(4f_2 - f_1)$. Another way, following the dark arrow, one obtains the new high-order harmonic frequencies of $2f_2$ and $3f_2$ by adding f_2 to itself and $2f_2$, respectively. In order to receive the third-order frequency $(2f_2 + f_1)$, one does the same way by adding f_2 to $(f_1 + f_2)$ in the direction of the dashed arrow. Similarly, the USB frequencies are generated in the same way.

References

1. Fahy, F., Gardonio, P.: Sound and Structural Vibration, 2nd edn, Academic Press, London (2007)
2. Crocker, M. J.: Noise and Vibration Control, Wiley, NY (2007)
3. Dawkins, S.: A guide to aeroacoustics: An overview, Lighthill's equation, related model equations, Webster's Digital Services (2011)
4. Schobeiri, M.: Turbomachinery Flow Physics and Dynamic Performance, 2nd edn, Springer, Berlin (2012)
5. Nguyen-Schäfer, H.: Rotordynamics of Automotive Turbochargers. Springer, Berlin (2012)
6. Nguyen-Schäfer, H. Kleinschmidt, R.: Analysis and nonlinear rotordynamics computation of constant tone in automotive turbochargers. 17th ATK Conference, Dresden (2012)
7. Tam, C.: Computational Aeroacoustics: A Wave Number Approach, Cambridge University Press, Cambridge (2012)
8. LMS Virtual Lab Acoustics: http://www.lmsintl.com/acoustic-simulation (2012)
9. Ehrich, F.: Handbook of Rotordynamics, Krieger Publishing Company, Florida (2004)
10. Software MSC: Program ASTRAN of Aero and Vibroacoustics (2012) http://www.mscsoftware.com/Products/CAE-Tools/Actran-For-Nastran.aspx

Chapter 4
Analyzing Root Causes of Noise

The induced noise types of automotive turbochargers in passenger vehicles have been already displayed in Chap. 2. Aerodynamic noise types, such as pulsation, rotational, growling, and whining noise, are generated from airflows in compressors of turbochargers by interactions between the fluid (charge air), solid surface (turbocharger), and fluid (ambient air) at their interfaces. Rotordynamic noise types like unbalance whistle, constant tone, high-order harmonic noise, and wear noise are created from rotordynamics of turbochargers by interactions between the fluid (oil), solid surface (turbocharger), and fluid (ambient air) at their interfaces. The unbalance whistle is generated by the rotor unbalance due to production tolerances; the constant tone is induced by the inner oil whirl occurring in the oil film bearings due to self-excitation instability. To improve the noise behavior of turbochargers in the vehicles, the root causes of these noise types should be analyzed in the following sections.

4.1 Aerodynamic Noise

Pulsation whistle, rotation, growling, and whining noise are the typical aerodynamic noise types in automotive turbochargers.

4.1.1 Root Causes of Aerodynamic Noise

Pulsation whistle is caused by slight differences in the compressor wheel (CW) chamber volumes due to milling or molding process variations. Due to the different chamber volumes, charge air pressures are slightly different at the blade outlets, as shown in Fig. 4.1. The generated charge air pressures in the CW chambers are substituted at the compressor outlet in every rotation cycle. This pressure perturbation δp causes the synchronous pulsation whistle that is

H. Nguyen-Schäfer, *Aero and Vibroacoustics of Automotive Turbochargers*,
DOI: 10.1007/978-3-642-35070-2_4, © Springer-Verlag Berlin Heidelberg 2013

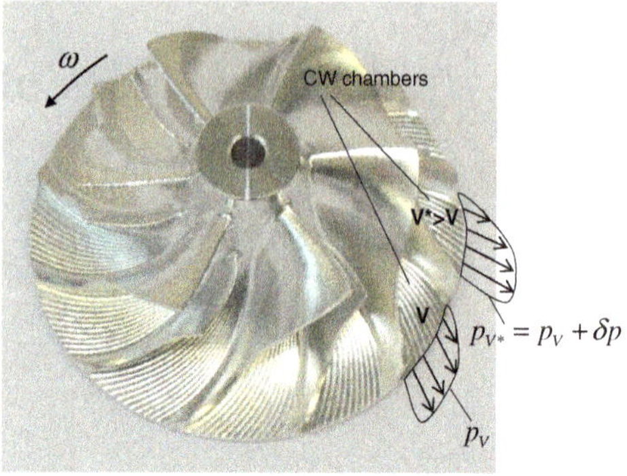

Fig. 4.1 Pressure distributions at the compressor wheel (*CW*) outlet

propagated through the compressor housing, charge air intercooler, and charge air duct. As a result, the charge air system is excited by the synchronous pulsation whistle, and radiates the pulsation whistle as airborne noise into the car cabin and environment. To reduce the pulsation whistle, a pulsation damper based on the Helmholtz's resonator is installed behind the compressor outlet in the charge air duct. Thus, the pulsation whistle amplitude is reduced in the damper.

Rotational noise is generated by rotating blades of the compressor or turbine wheels. In the case of a radial compressor, the charge air pressure is built in the compressor wheel chambers due to rotation. The rotating blades disturb the charge air pressure zone at the blade inlet, blade outlet, and near the compressor tongue in each rotation cycle. This causes air wakes and vortexes behind the rotating blades and near the compressor tongue, leading to local pressure disturbances that induce rotational noise in the turbocharger. The frequency of the rotational noise f_{RN} equals the product of the CW blades Z and rotor frequency $f_R = \omega/(2\pi)$, also $f_{\mathrm{RN}} = f_R.Z$. At the compressor inlet, Z equals the number of the main blades ($Z = 6$); at the compressor outlet, Z equals the number of the main and splitter blades ($Z = 12$), as displayed in Fig. 4.1.

The Waterfall plot measured at the compressor housing indicates that the rotational noise has sixth frequency order (6X) at the compressor inlet and 12th frequency order (12X) at the compressor outlet, as shown in Fig. 4.2. Generally, either a large number of the CW blades or an optimized geometry of the compressor tongue is used to reduce the rotational noise. The first one increases the noise frequencies (6 and 12X) above 16 kHz so that adult human being ears cannot recognize them, but animals could hear such noises at high frequencies. The optimized tongue at the compressor outlet reduces pressure disturbances (e.g., air wakes and vortexes) at the tongue. Therefore, the charge air pressure smoothly flows into the compressor outlet in each rotation cycle, causing little rotational noise.

Growling noise (compressor stall-related noise) is induced by the partial reversal of the charge air at the suction side near the blade outlet.

Whining noise (compressor surge-related noise) is produced by the deep surge condition in the compressor where the charge air totally recirculates from the

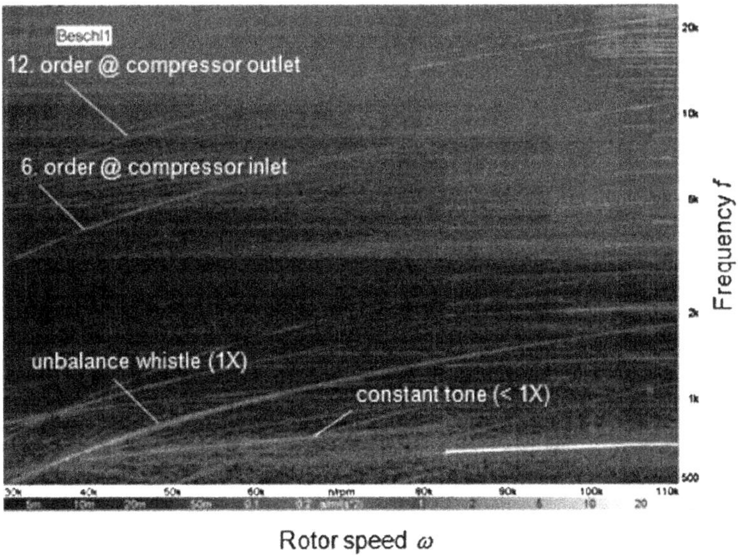

Fig. 4.2 Waterfall plot of noise spectra in turbochargers

compressor outlet to compressor inlet. The whining noise occurs when car driver suddenly releases gas pedal to decelerate the vehicle. At the moment, the currently required charge air mass flow rate for the engine, still at high turbocharger speed, is strongly reduced because the engine throttle valve begins to close. It leads to the pressure increase at the compressor outlet due to the still high turbocharger speed. This causes the deep surge working condition in the compressor, leading to whining noise in the turbocharger.

To better comprehend the root causes of the compressor rotating stall and surge, flow behaviors in the compressor must be studied and analyzed in the following subsections.

4.1.2 Aerodynamics of Radial Compressors

Figure 4.3 shows a backward swept radial compressor wheel that is mostly used in automotive turbochargers in order to improve the compressor performance map width. Note that the rotating direction ω of the CW is against the blade outlet direction; i.e., the outflow direction of the radial CW is against the circumferential velocity at the blade outlet.

The radial CW consists of main and splitter blades that divide the CW chambers into two equal chambers. The splitter blade has two functions: first, avoiding choke flow at the blade inlet 1; second, reducing flow separation at the suction side near the blade outlet 2. However, splitter blades cause more flow friction in the CW and possibly decreasing mass flow rate in the compressor.

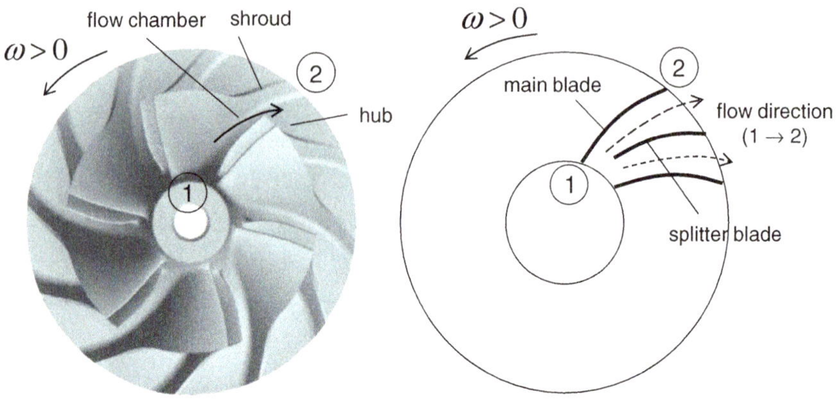

Fig. 4.3 Backward swept radial compressor wheel (*CW*)

Shroud is the contour on the top of the blades where the air gap is between the shroud and compressor housing. Over the shroud the airflow could leak from the pressure side to the suction side of the blades during rotating. Hub is the bottom surface (conical shape) located between the main and splitter blades from the blade inlet 1 to the blade outlet 2, as indicated in Fig. 4.3. Three-dimensional flows in the CW are limited in the flow pattern between the main and splitter blades, hub, shroud, and as well as compressor housing.

The compressor setup of a typical automotive turbocharger is displayed in Fig. 4.4. The compressor contains the CW with the super back, diffuser, volute, and compressor housing. Intake air enters the blade inlet 1 and is compressed along the blades from the compressor inlet 1 to the compressor outlet 2, and from the hub to shroud of the impeller (blade). Due to high rotational speed of the CW, the intake air pressure is increased at the blade outlet 2. The charge air pressure is

Fig. 4.4 Radial compressor setup

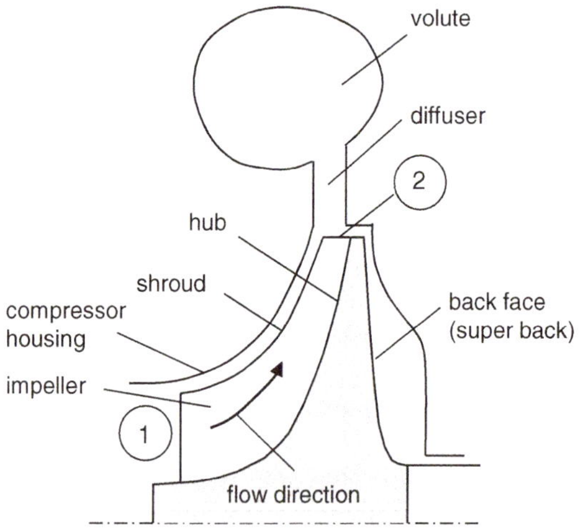

additionally gained in the diffuser built in the compressor housing between the CW and the volute. The charge air mass flow rate through the compressor, not the charge air pressure, is used for turbocharging downsized engines.

The compressor torque is derived from the angular momentum equation.

$$M_C = \dot{m}_C(r_2 c_{u,2} - r_1 c_{u,1}) \tag{4.1a}$$

where

$\dot{m}_C$ is the compressor mass flow rate;
r_1 and r_2 are the inlet (inducer) and outlet (exducer) radii of CW;
$c_{u,1}$ and $c_{u,2}$ are the tangential (or whirl) velocities at the inlet and outlet.

The compressor mass flow rate is resulted from the air density ρ, meridional velocity component c_m, and compressor cross-sectional area A_C.

$$\dot{m}_C = \rho c_m A_C \tag{4.1b}$$

The compressor power is resulted from the compressor torque and angular rotor speed.

$$\begin{aligned} P_C &= M_C\omega = \dot{m}_C\omega\,(r_2 c_{u,2} - r_1 c_{u,1}) \\ &= \dot{m}_C(u_2 c_{u,2} - u_1 c_{u,1}) \end{aligned} \tag{4.2a}$$

within

u_1 and u_2 are the circumferential velocities at the CW inlet and outlet.

The compressor power is calculated from the turbine power P_T and mechanical efficiency η_m.

$$P_C = \eta_m P_T = \eta_m \eta_T \dot{m}_T c_{p,g} T_3 \left[1 - \left(\frac{p_4}{p_3}\right)^{\left(\frac{\kappa-1}{\kappa}\right)_g}\right] \tag{4.2b}$$

where

η_T is the turbine efficiency;
T_3 is the exhaust gas inlet temperature;
p_3 is the exhaust gas inlet absolute pressure;
p_4 is the exhaust gas outlet absolute pressure;
κ_g is the exhaust gas isentropic coefficient ($\kappa_g \approx 1.32$);
$c_{p,g}$ is the exhaust gas heat capacity at constant pressure.

The required compressor power for the charge air compression is written in a function of the mass flow rate, inlet temperature, and pressure ratio of the compressor.

$$P_C = \frac{P_{C,ideal}}{\eta_C} \equiv \frac{\dot{m}_C c_{p,a} T_1}{\eta_C} \left[\left(\frac{p_2}{p_1}\right)^{\left(\frac{\kappa-1}{\kappa}\right)_a} - 1\right] \tag{4.2c}$$

where

η_C is the compressor efficiency;
T_1 is the air inlet temperature;
p_1 is the air inlet absolute pressure;
p_2 is the charge air outlet absolute pressure;
κ_a is the charge air isentropic coefficient ($\kappa_a \approx 1.4$);
$c_{p,a}$ is the charge air heat capacity at constant pressure.

Having combined Eqs. (4.2b) and (4.2c), one obtains the pressure ratio of the compressor π_C in the *first turbocharger equation*.

$$\pi_C \equiv \frac{p_2}{p_1} = \left(1 + \left(\frac{\dot{m}_T}{\dot{m}_C}\frac{T_3}{T_1}\eta_{TC}\right)\frac{c_{p,g}}{c_{p,a}}\left[1 - \left(\frac{p_3}{p_4}\right)^{-\left(\frac{\kappa-1}{\kappa}\right)_g}\right]\right)^{\left(\frac{\kappa}{\kappa-1}\right)_a} \tag{4.2d}$$

Within the overall turbocharger efficiency is defined by

$$\eta_{TC} = \eta_m \eta_T \eta_C \tag{4.2e}$$

Convention for the flow and blade angles of α_i and β_i:

- $\alpha_i > 0$ if $c_{u,i} > 0$; $c_{u,i}$ has the same direction of the circumferential velocity u_i;
- $\beta_i < 0$ if $w_{u,i} < 0$; $w_{u,i}$ is opposite to the circumferential velocity u_i.

The absolute gas velocity in the CW is the sum of the relative and circumferential gas velocities w_i and u_i, respectively, as shown in Fig. 4.5

$$\vec{c}_i = \vec{w}_i + \vec{u}_i \tag{4.3}$$

Therefore, the relative velocities are calculated for the CW inlet and outlet for $i = 1, 2$.

$$\begin{aligned} w_i^2 &= c_i^2 + u_i^2 - 2\vec{c}_i\vec{u}_i = c_i^2 + u_i^2 - 2c_i u_i \cos(\vec{c}_i, \vec{u}_i) \\ &= c_i^2 + u_i^2 - 2c_i u_i \cos\left(\frac{\pi}{2} - \alpha_i\right) = c_i^2 + u_i^2 - 2u_i c_i \sin\alpha_i \end{aligned} \tag{4.4}$$

The tangential vector component $c_{u,i}$ for $i = 1; 2$ are calculated from the velocity triangle map, as given in Fig. 4.5.

$$c_{u,i} = c_i \sin\alpha_i \tag{4.5}$$

Thus, the relative velocity components w_i for $i = 1; 2$ are rewritten by substituting Eq. (4.5) into Eq. (4.4) in

$$|w_i| = \sqrt{c_i^2 + u_i^2 - 2u_i c_{u,i}} \tag{4.6}$$

Having combined Eqs. (4.2a) and (4.6), one obtains the compressor specific work W_C in the Euler's turbocharger equation. This specific work W_C is defined by the compressor power per unit mass flow rate in the unit of W/(kg.s^{-1}) or J/kg.

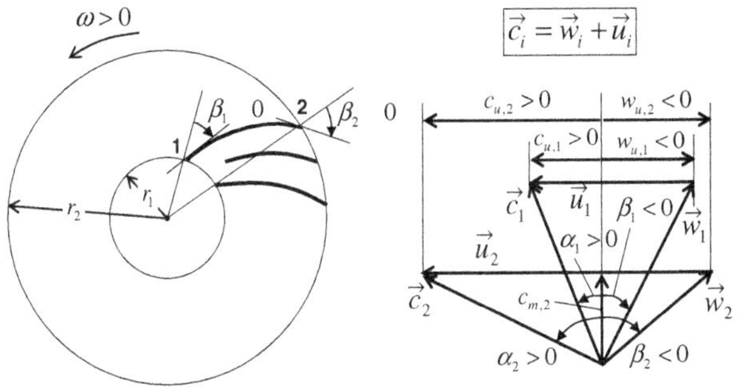

Fig. 4.5 Velocity *triangle* map of a radial compressor wheel (*CW*)

The Euler's turbocharger equation is formulated in

$$\begin{aligned} W_C &\equiv \frac{P_C}{\dot{m}_C} = (u_2 c_{u,2} - u_1 c_{u,1}) \\ &= \frac{1}{2}\left[(c_2^2 - c_1^2) + (u_2^2 - u_1^2) - (w_2^2 - w_1^2)\right] \end{aligned} \tag{4.7}$$

where

$w_2 < w_1$ because the outlet air density is larger than the inlet air density in case of compressors.

Note that to avoid the flow separation at the suction side on the compressor blades, the velocity ratio (w_2/w_1) should lie between 0.65 and 0.75 if the compressor performance is fulfilled according to De Haller's rule [1, 8].

The tangential component $c_{u,1}$ of the inlet absolute velocity is computed from the inlet absolute velocity c_1, and its inlet flow angle α_1 $(\alpha_1 > 0)$, as given in Fig. 4.5.

$$c_{u,1} = c_1 \sin \alpha_1 \tag{4.8}$$

The tangential component $c_{u,2}$ of the outlet absolute velocity is resulted from the outlet circumferential velocity u_2, meridional velocity component $c_{m,2}$, and outlet blade angle β_2 $(\beta_2 < 0)$, as shown in Fig. 4.5.

$$\begin{aligned} c_{u,2} &= c_2 \sin \alpha_2 \\ &= u_2 + c_{m,2} \tan \beta_2 \end{aligned} \tag{4.9}$$

Having applied the second law of thermodynamics for a steady-state adiabatic process, one obtains the compressor specific work

$$\begin{aligned} W_C &= u_2 c_{u,2} - u_1 c_{u,1} \\ &= h_{t,2} - h_{t,1} \\ &= c_p(T_2 - T_1) + \tfrac{1}{2}(c_2^2 - c_1^2) \end{aligned} \tag{4.10}$$

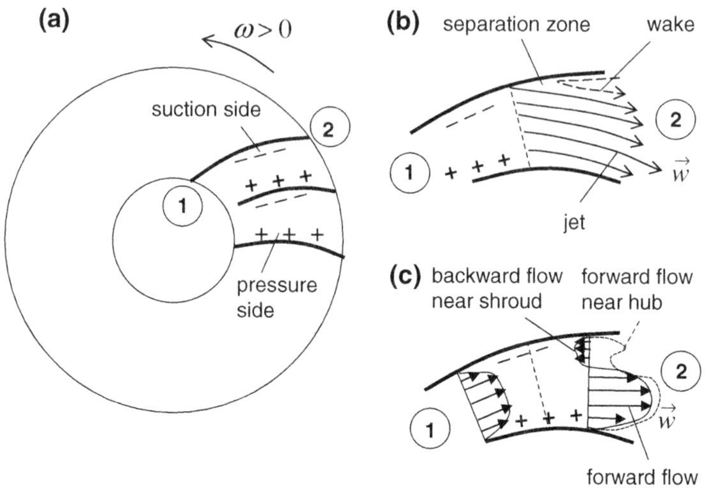

Fig. 4.6 Flow velocity profiles in a radial compressor wheel

where

$h_{t,1}$, $h_{t,2}$ total enthalpies at the CW inlet and outlet (see Appendix A);
c_p air heat capacity at constant pressure;
c absolute air velocity.

In the case of the backward swept CW mostly used in automotive turbochargers, the charge air is compressed on the impeller front face in the rotating direction ω, which is called the pressure side ($ps = + + +$), as indicated in Fig. 4.6a. On the contrary, the airflow separates from the impeller back face near the outlet 2, which is called the suction side ($ss = - - -$) opposite to the pressure side.

On the pressure side, the airflow jet occurs at the blade outlet 2 and remains on the blade with the forward flow during the rotation at a velocity ω (see Fig. 4.6b). On the suction side opposite to the pressure side, the charge air pressure decreases from the blade inlet 1 to the blade outlet 2. Therefore, the air wake begins separating from the suction side near the shroud where the relative velocity gradient at the wall equals zero. In the separation zone, the charge air partially recirculates at the negative velocity gradient at the wall; however, the volumetric flow rate in the CW is still positive with the forward flow. The air outflow is mostly forward even at the suction side near the hub. Note that the flow profile at the flow pattern outlet changes from the hub to shroud, as shown in Fig. 4.6c.

When the charge air partially separates from the suction side near the blade outlet, in the vaneless diffuser, or near the inducer tips of the CW, the *rotating stall* (also *mild surge*) happens (see Figs. 4.6 and 4.7). In this case, the mass flow rate is reduced but still positive in the compressor, causing however the *growling noise*.

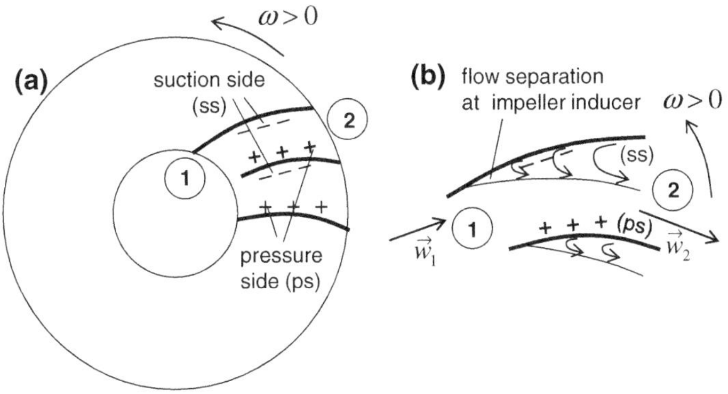

Fig. 4.7 Flow reversal in the impeller inducer

In contrast to the rotating stall, the charge air totally recirculates to the compressor outlet to the compressor inlet, from the suction side to the pressure side, and from the compressor hub to the compressor shroud at low air mass flow rates and high pressures, as shown in Fig. 4.8.

This phenomenon is called the compressor deep surge that mostly follows the rotating stall in case the pressure at the compressor outlet is much higher than the charge air pressure. In the deep surge, the mass flow rate takes turns periodically changing from the positive (forward flow) to negative (backward flow) mass flow rates. During the periodic airflow reversal (recirculation), very loud whining noise takes place in the compressor. Due to friction in the airflow recirculation, air temperature is strongly increased in the CW. Moreover, the deep surge causes damage the thrust bearing and turbocharger as well. Note that the compressor deep surge could occur in the diffuser when the pressure at the compressor outlet is relatively high at low mass flow rates so that the flow recirculates from the diffuser to the compressor inlet. In the deep surge, the compressor does not transport any

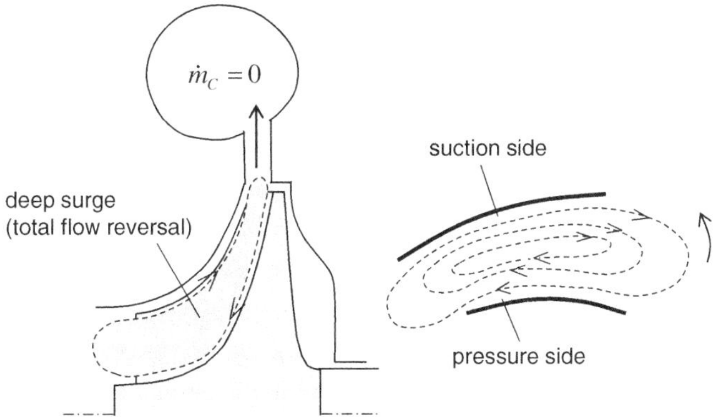

Fig. 4.8 Flow reversal in a radial compressor wheel

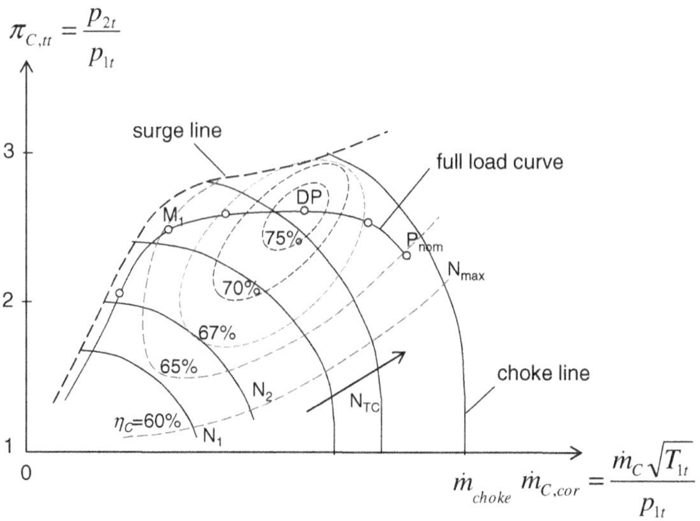

Fig. 4.9 Compressor performance map

mass flow rate at all because the airflow entirely recirculates from the compressor outlet to the compressor inlet and vice versa, as shown in Fig. 4.8.

These phenomena of the compressor rotating stall and deep surge will be analytically analyzed in the following subsection.

4.1.3 Stall and Surge in Radial Compressors

Flow instabilities in the compressor are to blame for rotating stall and surge mostly in the working condition at small mass flow rates and high pressures. The flow instabilities cause flow reversals that happen at certain operating points left from the surge line in the compressor performance map. The compressor ratio π_C versus corrected mass flow rate is displayed in the performance map (see Fig. 4.9). At low rotor speeds from the idle condition (about 30 % of N_{max}), the compressor pressure ratio increases with a relatively high pressure gradient in the full load curve to ensure a good transient response in low-end torque (LET). At further increasing the rotor speed N to nearly 70 % of N_{max}, the compressor pressure ratio increases to approximately 2.5, and the engine achieves the maximum torque M_1. The nominal power P_{nom} occurs just before the maximum operating rotor speed N_{max}. The maximum rotor speed corresponding with $u_{2,max}$ = 560 m/s is reached in highland applications where the rotor speed increases by 8–10 % (highland speed reserve) above the nominal operating rotor speed.

The compressor pressure ratio is formulated in a function of the corrected mass flow rate and turbocharger speed in the performance map.

$$\pi_C \equiv \frac{p_2}{p_1} = f(\dot{m}_C, N_{TC}) \tag{4.11}$$

where

π_C compressor pressure ratio;
p_1, p_2 air pressures at the compressor inlet and outlet;
$\dot{m}_C$ mass flow rates in the compressor.

At decreasing compressor volumetric flow rates, the charge air pressure rises in the turbocharger isospeed N_{TC}, as shown in Fig. 4.10. As the turbocharger isospeed passes the surge line, the compressor rotating stall occurs where the charge air partly recirculates in the CW periodically. In the stall condition, the compressor pressure ratio decreases with reducing volumetric flow rates; however, the volumetric flow rates in the compressor are still positive. The charge air partly recirculates from the compressor outlet to the compressor inlet only in a short time and is pumped again to the compressor outlet; thus, the rotating stall period is relatively short. In contrast, in the deep surge condition the compressor volumetric flow rate decreases in the direction of negative volumetric flow rates at further reducing compressor pressure ratios. Suddenly, the duct pressure behind the compressor drops below the compressor inlet pressure; thus, the charge air is compressed again to the compressor outlet in a normal condition on the rhs of the surge line (see Fig. 4.10). In this case, the surge period lasts much longer than the rotating stall period because the large duct volume behind the compressor must be completely discharged to a minimum pressure that is below the current compressor inlet pressure. Then, the duct is charged again with the compressed charge air; its pressure increases again above the entering charge air pressure, leading to compressor stall and surge. As a result, the surge process is periodically repeated in the

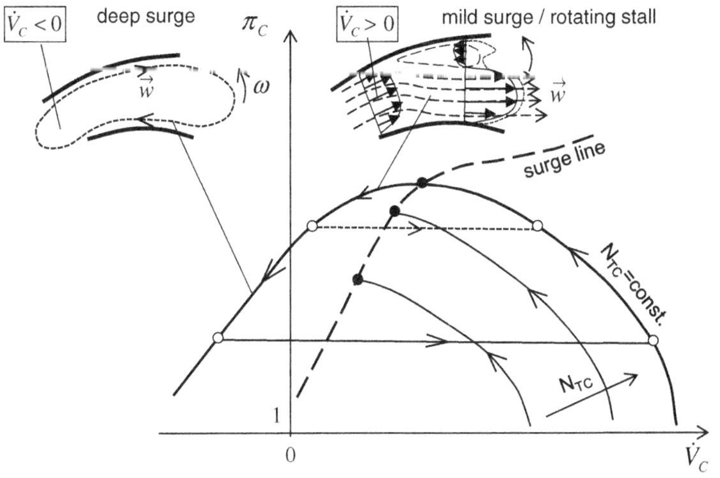

Fig. 4.10 Compressor stall and surge in the performance map

surge working condition. Both rotating stall and surge flow conditions cause the very disturbing growling and whining noises, respectively for driver and passengers in the vehicle cabin. Additionally, the charge air temperature is drastically increased in the compressor due to repeating periodic flow reversals. In this case, the thrust load acting on the thrust bearing takes turns changing from the turbine wheel to CW direction and vice versa due to the very large pressure pulsations of the charge air, leading to damage of the thrust bearing and the whole turbocharger as well.

Note that compressor surge has been considered in one-dimensional flow pattern; however, it could be turned into the three-dimensional rotating stall instability, as shown in Fig. 4.8. At first, mild surge occurs when the operating flow condition in the turbocharger isospeed passing the surge line at reducing volumetric flow rates. There begins the airflow to separate from the suction side near the blade outlet. Usually, the deep surge follows the mild surge (rotating stall) directly when the entire flow reversal takes place in the compressor. However, deep surge and rotating stall could take turns occurring in certain operating conditions.

To improve the flow stabilities for the stall and surge behaviors in radial compressors, the outlet blade angle $|\beta_2|$ should be increased (i.e., more backward swept blades). However, the larger the backward swept blade angle is, the higher the flow resistance exists in the CW, causing the compressor efficiency reduction. Additionally, the normal stresses caused by the centrifugal force acting on the blades also increase due to the longer blade length, leading to blade damage in some cases.

In the following section, the thresholds of instability of the compressor rotating stall and surge are analyzed and discussed. At first, some useful dimensionless numbers and coefficients are defined for analyzing the compressor stall and surge instabilities in the performance map.

The flow coefficient ϕ is defined by the ratio of the meridional velocity to the circumferential velocity at the blade outlet 2.

$$\phi \equiv \frac{c_{m,2}}{u_2} \propto \frac{\dot{V}_C}{D_2^2 u_2} \propto \frac{\dot{m}_C}{\rho_2 D_2^2 u_2} \tag{4.12}$$

The pressure rise coefficient ψ is defined by the ratio of the pressure increase in the compressor to the kinetic energy at the blade outlet 2.

$$\psi \equiv \frac{\Delta p_C}{\rho_2 N_{TC}^2 D_2^2} \propto \frac{\Delta p_C}{\rho_2 u_2^2} \propto \frac{\pi_C}{\rho_2 u_2^2} \tag{4.13}$$

The compressor characteristic C is the behavior between the pressure rise and flow coefficients, as shown in Figs. 4.10 and 4.13. The throttle characteristic T of the compressor system containing the duct, plenum, and throttle (see Fig. 4.12) describes the volumetric flow rate in the compressor depending on the throttle coefficient α_T and the pressure drop Δp_T.

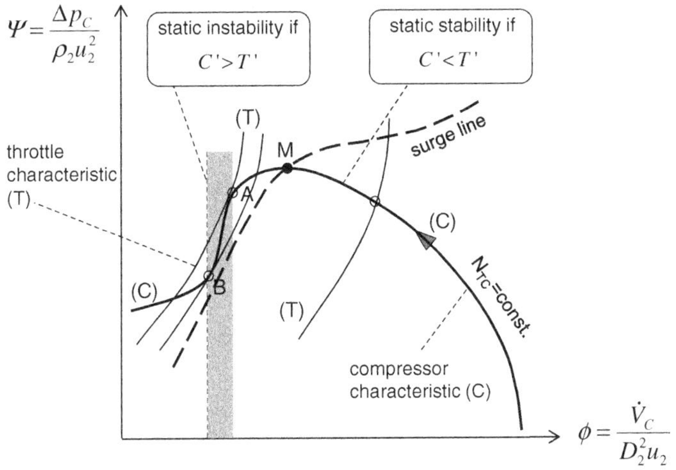

Fig. 4.11 Static instability operating condition

$$\dot{V}_T \propto \alpha_T A_T \sqrt{\Delta p_T} \tag{4.14a}$$

The relation between the flow and pressure rise coefficients is resulted from Eqs. (4.12), (4.13), and (4.14a).

$$\psi_T \propto \frac{1}{(A_T \alpha_T)^2} \phi_T^2 \tag{4.14b}$$

Figure 4.11 shows that the stable condition (without stall and surge) locates on the rhs of the surge line. Left from the surge line, the compressor characteristic is statically stable between M and A in the isospeed N_{TC}. Then, it becomes statically unstable between A and B when the compressor characteristic slope C' is larger than the throttle characteristic slope T' according to [3].

$$\left(\frac{\partial \psi}{\partial \phi}\right)_C \equiv C' > T' \equiv \left(\frac{\partial \psi}{\partial \phi}\right)_T \tag{4.15}$$

By differentiating Eq. (4.14b), one obtains the throttle characteristic slope T' that is always positive.

$$T' \equiv \left(\frac{\partial \psi}{\partial \phi}\right)_T \propto \frac{2\phi_T}{(A_T \alpha_T)^2} = \frac{2\xi_T^2 \phi_T}{A_T^2} > 0 \tag{4.16}$$

where

α_T is the throttle coefficient;

ξ_T is the pressure loss coefficient of the throttle ($\xi_T \equiv 1/\alpha_T$).

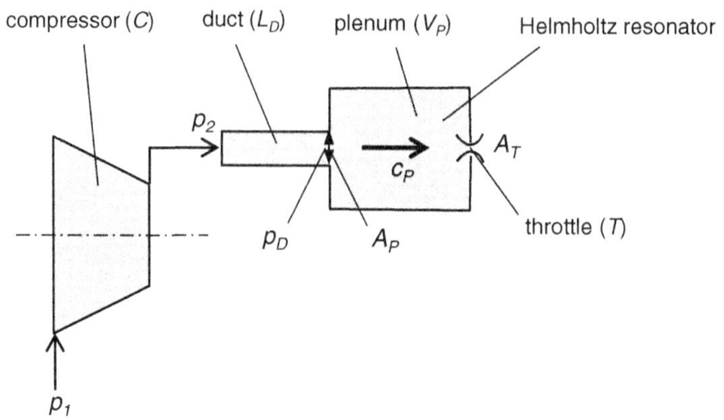

Fig. 4.12 Compressor system setup

Equation (4.16) indicates that the throttle characteristic slope T' is proportional to the throttle flow coefficient ϕ_T. Therefore, the lower the throttle flow coefficient ϕ_T, the smaller the throttle characteristic slope T' is. The volumetric flow rate in the throttle is limited by the throttle area A_T (see Fig. 4.12). If the compressor volumetric flow rate ($\phi_C \sim C'$) is larger than the maximum allowed volumetric flow rate in the throttle ($\phi_T \sim T'$), the airflow reverses in the compressor. In case of $C' > T'$, as given in Eq. (4.15), the static flow instability takes place in the compressor. In this case, the airflow separates from the suction side near the blade outlet and recirculates to the compressor inlet, leading to compressor rotating stall and deep surge.

Logically, the compressor working condition is statically stable only if

$$C' < T' \tag{4.17}$$

In this case, the compressor volumetric flow rate is less than the maximum allowed volumetric flow rate in the throttle; thus, the compressor operates in a statically stable condition and satisfies the stability condition given in Eq. (4.17). Both statically unstable and stable conditions are displayed in the compressor performance map, as shown in Fig. 4.11.

The threshold of static instability (onset of static instability) takes place if the compressor characteristic slope C' equals the throttle characteristic slope T'. In this working condition, the airflow begins separating from the suction side near the blade outlet.

$$C' = T' \tag{4.18}$$

At the *static instability*, as given in Eq. (4.15), the airflow collapses and separates from the blade. The limited small volumetric flow rate in the throttle is to blame for the static instability. On the contrary, the negative damping of the airflow vibration in the compressor system causes *dynamic instability* [1]. In the following section, the dynamic instability behavior of the compressor is thoroughly analyzed and discussed.

Figure 4.12 shows the compressor system containing the compressor C, duct with a length L_D and cross-sectional area A_P, plenum with a volume V_P and sound velocity c_P, and throttle T. Due to pressure reflection at the throttle, airflow vibrates in the compressor system with amplitude $\zeta(t)$ and angular frequency ω.

The homogenous vibration equation of the characteristics of the mass flow rate, perturbed pressure, and density is written for the compressor system.

$$M\ddot{\zeta}(t) + D\dot{\zeta}(t) + S\zeta(t) = 0 \tag{4.19}$$

Within

$\zeta(t)$ is a perturbation characteristic,
M, D, and S are the vibration equation coefficients computed from the flow rate equations of the compressor system [1].

The system-mass coefficient M is defined by

$$M \equiv \left(\frac{L_D V_P}{A_P c_P^2}\right) T' > 0 \tag{4.20}$$

The system-damping coefficient D is written in

$$D \equiv \left(\frac{L_D}{A_P} - \frac{C'T'V_P}{c_P^2}\right) \tag{4.21}$$

The system-stiffness coefficient S is given by

$$S \equiv (T' - C') \tag{4.22}$$

Equation (4.19) is a linear second-order ordinary differential equation (ODE) whose homogeneous solution results in

$$\begin{aligned} \zeta(t) &= Ae^{\lambda t} \\ &= Ae^{(\alpha+j\omega)t} = Ae^{\alpha t}e^{j\omega t} \\ &= Ae^{\alpha t}(\cos\omega t + j\sin\omega t) \end{aligned} \tag{4.23}$$

where
A is the vibration response amplitude of $\zeta(t)$;
λ is the complex eigenvalue of the vibration system ($\lambda = \alpha + j\omega$);
ω is the surge angular frequency;
j is the imaginary unit ($j^2 = -1$).

Note that the system response $\zeta(t)$ is conditionally stable if the real term α of the complex eigenvalue λ is negative; otherwise, at $\alpha > 0$ the solution exponentially increases with time. This leads to the system instability that causes rotating stall and surge. In the following section, the real term α (growth/decay rate) is computed.

Having substituted the first and second derivatives of $\zeta(t)$ into Eq. (4.19), one obtains the characteristic equation of the vibration system [4].

$$M\lambda^2 + D\lambda + S = 0 \tag{4.24}$$

The eigenvalue λ is resulted from solving Eq. (4.24).

$$\lambda = \alpha + j.0 = -\frac{D}{2M} \pm \sqrt{\left(\frac{D}{2M}\right)^2 - \left(\frac{S}{M}\right)} \tag{4.25}$$

The system-stiffness coefficient S in Eq. (4.25) must be positive so that the growth/decay rate α is always negative. Thus, $S > 0$ satisfies the condition of $\alpha < 0$ that is necessary for the compressor static stability of $C' < T'$.

However, Eq. (4.25) can be formulated in the complex number when the term in the square root is negative.

$$\lambda = \alpha \pm j\omega = -\frac{D}{2M} \pm \mathrm{j}\sqrt{\left(\frac{S}{M}\right) - \left(\frac{D}{2M}\right)^2} \tag{4.26}$$

Obviously, one obtains

$$\alpha = -\frac{D}{2M};\ \omega = \sqrt{\left(\frac{S}{M}\right) - \left(\frac{D}{2M}\right)^2} \tag{4.27}$$

To ensure that $\alpha < 0$, the system-damping coefficient D must be positive (see Appendix D); therefore, $D > 0$ is required for the compressor dynamic stability condition, as given in Eq. (4.29).

Figure 4.13 shows the stability behaviors of vibration responses at different values of α. The rotor response is stable at $\alpha < 0$ where the compressor surge amplitude exponentially dies down with time. The threshold of instability takes place at $\alpha = 0$, where the vibration response amplitude is constant over time. Finally, the vibration response is unstable (compressor surge condition) at $\alpha > 0$, where the surge amplitude exponentially increases with time.

Alternatively, the Routh-Hurwitz stability criterion can be used to analyze the flow stability of the compressor system vibration (see Appendix E) [4]. The airflow behavior is conditionally stable if all coefficients M, D, and S in Eq. (4.24) are nonzero and have the same positive or negative signs to satisfy the necessary conditions for stability. First, the first coefficient M is positive because the throttle characteristic slope T' is always positive. Second, the positive system-stiffness coefficient $S > 0$ fulfills the static stability condition of $\mathrm{C}' < \mathrm{T}'$ at $\alpha < 0$.

$$\begin{aligned} S = (T' - C') &> 0 \\ \Rightarrow C' &< T' \end{aligned} \tag{4.28}$$

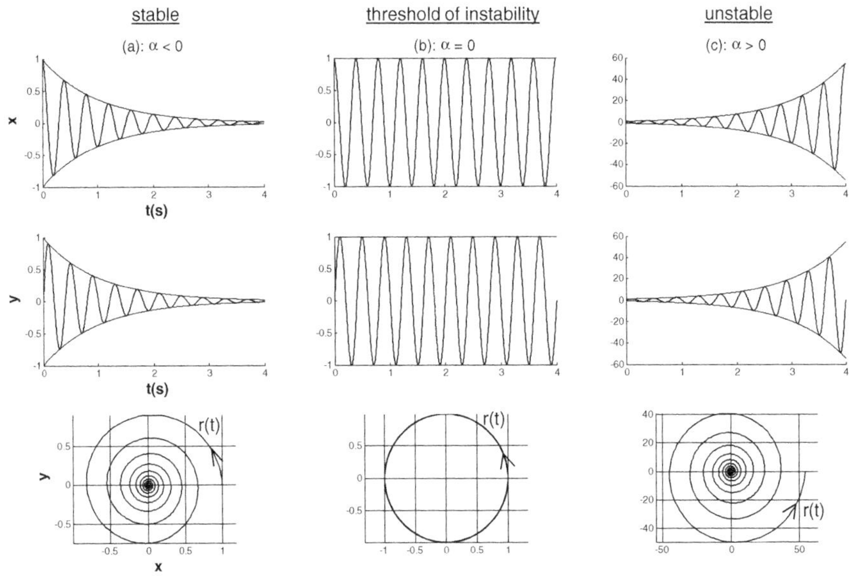

Fig. 4.13 Stability behaviors of vibration responses

The result of Eq. (4.28) indicates that the static stability condition given in Eq. (4.17) has been reconfirmed using the Routh-Hurwitz criterion (q.e.d.).

Third, the dynamic stability condition of $\alpha < 0$ is satisfied if the system-damping coefficient D in Eq. (4.24) is positive.

$$D = \left(\frac{L_D}{A_P} - \frac{C'T'V_P}{c_P^2} \right) > 0$$
$$\Rightarrow C' < \frac{L_P c_P^2}{A_P V_P T'} \equiv \frac{1}{B'^2 T'} \tag{4.29}$$

where the factor B' is defined by

$$B'^2 \equiv \frac{A_P V_P}{L_P c_P^2} = \frac{1}{\omega_H^2} \left(\frac{A_P}{L_D} \right)^2; \quad B' \text{ in unit } [\text{m.s}]. \tag{4.30}$$

The Helmholtz frequency of the plenum/duct is defined by

$$\omega_H \equiv c_P \sqrt{\frac{A_P}{L_D V_P}}; \quad \omega_H \text{in unit } \left[\text{s}^{-1} \right] \tag{4.31}$$

The factor B' changes inversely related to the Helmholtz frequency squared. The Helmholtz frequency becomes small in case of the large plenum volume V_P, the long duct length L_D, and the small cross-sectional area A_P according to Eq. (4.31). Therefore, the factor B' is quite large at the small Helmholtz frequency. At the large factor B', the stability condition in Eq. (4.29) is not fulfilled any longer.

Therefore, the dynamic instability occurs in the compressor, leading to compressor surge.

The undamped eigenfrequency (natural frequency) of the compressor system is computed from Eq. (4.19).

$$\begin{aligned} \omega_n &\equiv \sqrt{\frac{S}{M}} = c_P \sqrt{\frac{A_P}{L_D V_P}} \cdot \sqrt{\frac{T' - C'}{T'}} \\ \Rightarrow \omega_n &= \omega_H \sqrt{1 - \frac{C'}{T'}} < \omega_H \end{aligned} \tag{4.32}$$

This result indicates that the Helmholtz frequency ω_H of the compressor system is normally higher than the surge frequency ω given in Eq. (4.27).

Therefore, the compressor working condition becomes dynamically unstable at $\alpha > 0$ in a negative damping system according to Eq. (4.29).

$$C' > \frac{1}{B'^2 T'} \tag{4.33}$$

The dynamic instability condition is much critical at large values B' and T' that lead to a very small rhs term in Eq. (4.33). In this case, the compressor characteristic slope C' easily exceeds this rhs term although the compressor characteristic slope C' is smaller than the throttle characteristic slope T', as shown in Fig. 4.14. As a result, the dynamic stability condition in Eq. (4.29) must be taken into account at first. Then, the static stability condition in Eq. (4.28) is considered.

Generally, the factor B' in unit [m s] is transformed into the dimensionless factor B [-] which was defined by Greitzer [3].

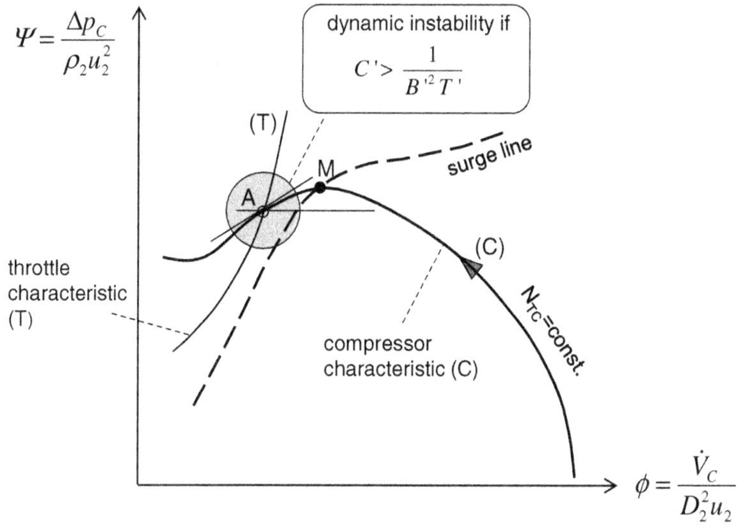

Fig. 4.14 Dynamic instability operating condition

$$\begin{aligned} B &\equiv B'.\left(\frac{u_2}{2A_P}\right) = \frac{u_2}{2\omega_H L_D} \\ &= \frac{u_2}{2c_P}\sqrt{\frac{V_P}{A_P L_D}} \end{aligned} \tag{4.34}$$

within

u_2 is the circumferential velocity at the impeller outlet;

c_P is the sound velocity in the plenum.

Experimental results have shown that the threshold of the dynamic instability (deep or mild surge) takes place in the compressor when the dimensionless factor B is between 0.7 and 0.8 [1].

4.2 Unbalance Whistle

4.2.1 Root Causes of Unbalance Whistle

Production process of CW and turbine shaft causes an initial unbalance, in which the mass center does not exactly locate in the geometrical axis of the CW or turbine shaft, leading to unbalance forces acting on the rotor. Besides the initial unbalance due to producing tolerances, additional rotor unbalances are resulted from misaligned mounting the CW in the turbine shaft. This misalignment generates an unbalance moment acting upon the rotor [4]. Excessively large unbalance forces and moments induce high response amplitudes; these lead to bearing wear, seizure of the journal in the bearings, and rub contact between the compressor and turbine wheels and their housings. Moreover, the rotor unbalance generates the unbalance whistle whose frequency is synchronous with the rotor frequency (i.e., 1X synchronous noise).

The unbalance force is proportional to the unbalance and rotor speed squared. This is a dipole noise force acting on the turbocharger and inducing the unbalance whistle. The unbalance whistle is one of the disturbing airborne noise types of automotive turbochargers. To reduce the unbalance whistle, the turbocharger rotor must be high-speed balanced at two balancing planes.

There are two rotor balancing types of automotive turbochargers:

- *Low-speed balancing* (*shop balancing*) is used for rigid rotors at a low balancing speed up to 3,400 rpm, depending on the balancing machine types, to reduce the initial unbalance caused by the production process. Generally, the whole rotor is not balanced at the shop balancing, but only the CW and turbine shaft are separately balanced at a rigid state with the low-speed balancing. Hence, it is also named single part balancing. The low-speed balancing is generally carried out in two balancing planes at the nose and back face of the wheel. The goal of the low-speed balancing is to reduce the unbalance excitations; therefore, this prevents bearing damage due to rub contact and seizure between the journal and radial bearings caused by an excessively unbalanced rotor.

- *High-speed balancing* (*trim balancing*) is applied to flexible rotors at high balancing speeds of nearly 200,000 rpm, generally above the first bending critical speed. At high rotor speeds, the rotor deflects in the radial direction, causing an additional static unbalance. The static and mounting-related unbalances must be removed by the trim balancing in order to improve aeroacoustic qualities of automotive turbochargers for passenger vehicles. The high-speed balancing is carried out in two balancing planes at the CW screw-nut and hub surface between the blades near the CW outlet. The trim balancing purpose is to reduce the unbalance whistle noise level in passenger vehicles. Due to balancing costs, the trim balancing has been usually omitted in turbochargers for commercial vehicles and industrial applications unless the customers optionally request.

4.2.2 Unbalance Types of Turbochargers

Manufacturing CW and turbine shaft (i.e., turbine wheel and rotor shaft) induces static, couple, and dynamic unbalances. The static unbalance occurs when the polar mass-inertia axis differs from the rotational axis by an eccentricity ε. On the contrary, the couple unbalance occurs when the polar mass-inertia axis differs from the rotational axis by a misalignment angle α; however, they intersect each other at the mass center G.

In the static unbalance, the rotor mass center G always lies below the rotational axis in the equilibrium due to the gravity (see Fig. 4.15a). That means the rotor will return from any position above the rotational axis to the equilibrium position because of its weight. Hence, one calls it *static unbalance*; obviously, it is easily seen that the rotor moves itself to the equilibrium position. On the contrary, in the couple unbalance the rotor does not return to the equilibrium position because the

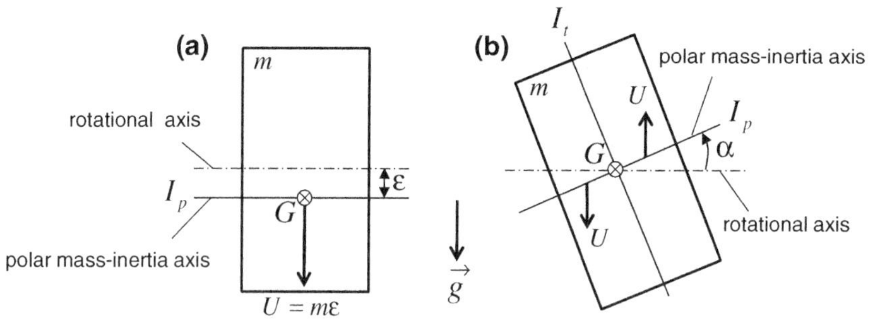

Fig. 4.15 **a** Rotor static unbalance. **b** Rotor couple unbalance

rotor mass center G is always in the rotational axis. The rotor couple unbalance is only recognized in the rotating condition in which the unbalance moment acts upon the rotor despite without static unbalance, as shown in Fig. 4.15b.

The static unbalance is caused by producing tolerances, such as large eccentricity at welding the turbine wheel on the rotor shaft, nonhomogeneous material, and residual unbalance after the shop balancing. Moreover, it is also induced by unbalance changes like thermal deformations, wear, and CW plastic deformations generated by rotor over-speeding, overheating the turbine wheel, and hard particle impact on the CW in low-pressure exhaust gas recirculation (EGR) turbocharging. On the contrary, the couple unbalance is created by a misalignment angle at welding turbine wheel in the rotor shaft, and unbalance changes, such as thermal and plastic deformations of the rotor due to rotor over-speeding, or loose screw-nut of the CW in the operation. In practice, both static and couple unbalances mostly occur in the turbine shaft manufacture, and CW assembly in the rotor shaft; these lead to the rotor dynamic unbalance.

The dynamic unbalance is similar to the couple unbalance, but the rotor mass center G does not lie in the rotational axis (see Fig. 4.16). Therefore, the dynamic unbalance can be decomposed in the static and couple unbalances. In fact, neither the static unbalance nor the couple unbalance exists alone in automotive turbochargers; instead, the dynamic unbalance mostly occurs. The dynamic unbalance is generated in the turbine shaft manufacture at welding the turbine wheel in the rotor shaft. Additionally, the roundness and manufacturing quality of the screw-thread shape play a key role for causing the dynamic unbalances. The eccentricity ε causes the static unbalance; the misalignment angle α induces the couple unbalance. Both unbalances are combined to the dynamic unbalance, as displayed in Fig. 4.16. The turbine shaft unbalance is the largest part of the entire rotor unbalance in the production. It contains the initial unbalance of the turbine wheel itself, eccentricity and misalignment at welding the turbine wheel in the rotor shaft.

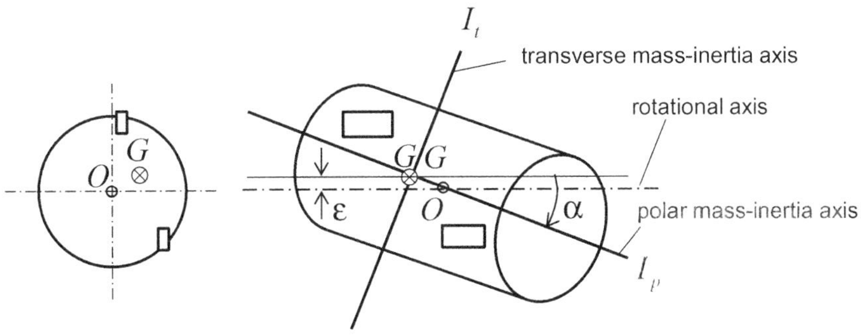

Fig. 4.16 Rotor dynamic unbalance

4.2.3 Unbalance Change

The rotor dynamic change after a long-term operating time is considered as the unbalance change. Thermal and plastic deformations or location changes of the compressor and turbine wheels are to blame for the unbalance change. The maximum unbalance change is approximately 100 % of the delivery acceleration responses after the trim balancing according to experience. Therefore, the long-term changed unbalance of the rotor increases to nearly twice the delivery acceleration responses. About one-third of the unbalance change goes to the turbine shaft due to heat-related plastic deformations, internal stress relaxations at the welding interface between the rotor shaft and turbine wheel; two-third is caused by changing the CW positions in the axial, radial, and rotational directions. The CW unbalance change is generated by its new inappropriate positions of eccentricity, misalignment, rotating about the turbine shaft. These are caused by the wheel plastic deformation, the wheel dislocation in the radial direction due to over-speeding, and the wheel misalignment in the rotor shaft because of loose screw-nut at the compressor inlet.

As a result, the *delivery acceleration levels* at the trim balancing must be reduced by at least one-half of the *limit acceleration levels* in which the unbalance whistle is inaudible in the vehicle cabin. The unbalance change ratio between the limit and delivery acceleration responses must be statistically determined by testing many turbochargers of the same type for the same vehicle type in a long-term operating time. If the ratio is chosen too large, many balancing steps are required for the trim balancing to keep the delivery acceleration levels small, leading to much time needed for the balancing process. Furthermore, this causes additional costs of lowered producing cycles, and cutting tools wear; therefore, more balancing machines are required for a large mass production. Otherwise, if the unbalance change ratio is selected too small, the long-term changed unbalance of the rotor exceeds the limit acceleration levels after a long-term operating time. As a result, the unbalance whistle is audible again in the cabin due to the unbalance change.

4.2.4 ICM Trim Balancing

The ICM trim balancing is based on the Influence Coefficient Method, which has been developed by Goodman, Rieger, Lund and Tonnesen, Tessarzik, Badgley, and Anderson, etc., since 1961 [5, 6]. To improve the aeroacoustic quality (NVH) of automotive turbochargers, the trim balancing is normally carried out in two balancing planes at the screw-nut (plane 1*) and the hub surface between the blades near the CW outlet (plane 2*), as shown in Fig. 4.17.

Based on the complex transfer impedance Z of the balancing system, i.e., the inversion of the complex dynamic stiffness coefficient K_S of the system, including

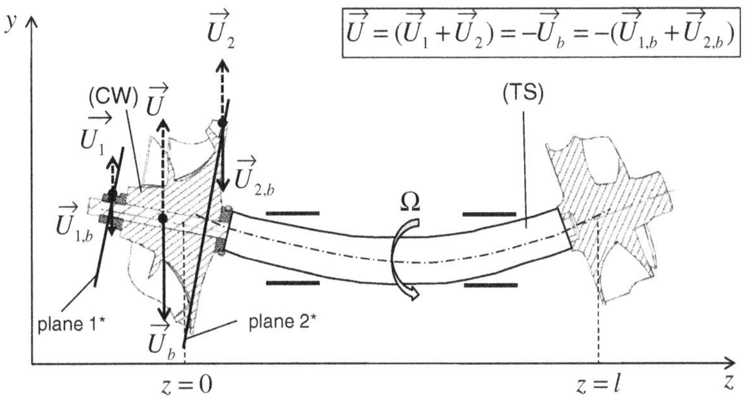

Fig. 4.17 ICM trim balancing at the balancing planes 1* and 2*

the turbocharger rotor and less influent balancing machine, the unbalance response at trim balancing is resulted from the complex transfer impedance and unbalance excitation force. At the trim balancing, note that *to let the turbocharger possibly free vibrate in the test rig; do not clamp it on the test bank*!

The measured unbalance responses could be rotor vibration velocities or accelerations measured in the measuring planes. Theoretically, the measuring planes must be the same positions of the balancing planes 1* and 2* in the CW. However, the balancing planes rotate at high rotor speeds. It is cost intensive to measure the rotor unbalance responses in the rotating balancing planes in the mass production. In practice, they are computed from the measured rotor responses in the certain nonrotating position in the bearing housing or compressor and turbine casings at the trim balancing. This nonrotating position must be experimentally determined depending on turbocharger types.

Due to the unbalance whistle, the rotor must be balanced at the high-speed trim balancing. Moreover, the rotor is flexible at high rotor speeds; it causes additional unbalances and unbalance moments. For this reason, the rotor should be balanced at high speeds in two balancing planes in the balancing plane 1* at the screw-nut and balancing plane 2* at the hub surface.

A nozzle ring supplied by pressurized air at about 0.3 MPa absolute is used to accelerate the rotor up to the balancing speed of about 200,000 rpm. The pressurized air energy is transformed into kinetic energy in the nozzle ring with an air jet at the sonic speed acting upon the turbine wheel. Therefore, the turbine wheel is propelled by the nozzle ring at the wheel inlet.

By twice timely differentiating the rotor amplitude, one obtains the acceleration amplitude measured in the trim balancing.

$$|a| = |\ddot{y}(t)| = \Omega^2 |y(t)| \propto N^2 |y(t)| \tag{4.35}$$

Note that the measured acceleration amplitude is proportional not only to the rotor amplitude but also to the rotor speed squared. It displays not exactly the rotor

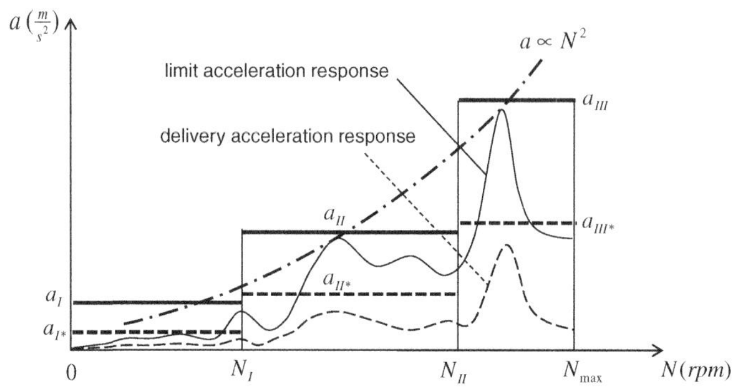

Fig. 4.18 Limit and delivery acceleration levels at the ICM trim balancing

amplitude behavior, but the measured acceleration is strongly affected by the rotor speed squared N^2, as shown in Fig. 4.18. As discussed in Sect. 5.5, the typical linear resonance does not exist in nonlinear rotordynamics, but only the limit cycle of the rotor response occurs at each rotor speed. The unbalance amplitude (1X) is relatively small compared to the asynchronous amplitudes in the rotor speed range. According to Eq. (4.35), the acceleration amplitude at high rotor speeds (N between N_{II} and N_{max}) could be higher than the acceleration amplitude at the quasi-resonance (N_{res} between N_I and N_{II}) although its rotor amplitude is less than the resonance rotor amplitude. Therefore, the radial bearing failure due to rub contact at the trim balancing happens at the first quasi-resonance instead of the maximum acceleration amplitude at high rotor speeds. Figure 4.18 and Table 4.1 give the delivery and limit acceleration levels for the rotor acceleration responses in the ICM trim balancing process.

4.3 Subsynchronous Constant Tone

Inner oil whirl in the radial bearings of automotive turbochargers is to blame for the constant tone (howling), whose frequency ranges from 600 to 1,000 Hz in the human audible range. In rotating floating ring bearings, the inner oil whirl frequency order reduces from about 0.4 to 0.3X at increasing rotor speeds. Therefore,

Table 4.1 Delivery and limit acceleration levels in the balancing speed ranges

Balancing speed range N (rpm)	Delivery acceleration levels a_* (m/s^2)	Limit acceleration levels a (m/s^2)
$0 < N < N_I$	$a_{I*} < a_I$	a_I
$N_I \le N < N_{II}$	$a_{II*} < a_{II}$	a_{II}
$N_{II} \le N < N_{max}$	$a_{III*} < a_{III}$	a_{III}

the inner oil whirl frequency varies in small frequency range between 600 and 1,000 Hz compared to the rotor frequency; this frequency is usually considered as quasi-constant in the speed range of automotive turbochargers. The constant tone mostly occurs at the engine speed range between 1,500 and 3,500 rpm in second to fifth gear with middle to high loads.

The inner oil whirl is caused by nonlinear characteristics of the oil film bearings. It is a subsynchronous vibration response whose frequency is less than the rotor frequency. As long as the damping force in the bearing is larger than the destabilizing force caused by the bearing cross-coupled stiffness coefficient, the journal whirling amplitude still remains small in quasi-stable condition. Hence, the constant tone is inaudible in the vehicle cabin. The journal begins whirling in the bearing inner oil film when the destabilizing force exceeds the damping force at increasing rotor speeds, leading to the self-excitation instability (subsynchronous inner oil whirl) which generates the constant tone. On the contrary, the outer oil whirl does not induce any noise due to its low frequency. Note that the noise intensity is proportional to its frequency squared. The outer oil whirl frequency is relatively low compared to the inner oil whirl frequency; therefore, the noise intensity is negligibly small. As a result, the outer oil whirl is inaudible in the car cabin although its whirl amplitude is much larger than the inner oil whirl amplitude. The problematic constant tone will be thoroughly dealt with in Sect. 6, in which the mutual effects between the constant tone and nonlinear rotordynamics with two oil film bearings are analyzed and discussed.

4.4 High-Order Harmonic Noise

Ball bearings have no intrinsic damping due to very small clearances between the balls and races (see Fig. 4.20). As a result, they need an additional external damping to suppress the rotor amplitude responses at resonances and as well as to reduce airborne noises, such as the unbalance whistle and other aerodynamic noises induced by the ball bearings. In this case, a squeeze-film damper at the outer race is necessary for automotive turbochargers using ball and roller bearings. Contrary to the oil film bearings, the rolling element bearings do not induce the inner oil whirl; therefore, constant tone does not occur. However, besides the unbalance whistle (1X) the rolling element bearings have vibrations with multiple harmonic frequency orders of 2X, 3X, 4X, …, and modulation sidebands frequencies, as shown in Fig. 4.19.

The inappropriate setup mechanism of the bearing cartridge in the bearing center housing is to blame for the high-order harmonic and side band noises [4]. Due to the unsuitable damper oil film between the bearing cartridge and bearing housing, high-order harmonic noises are generated at multiple harmonic frequencies. At small damper oil film thickness, induced noise is not suppressed enough and therefore transferred through the bearing center housing.

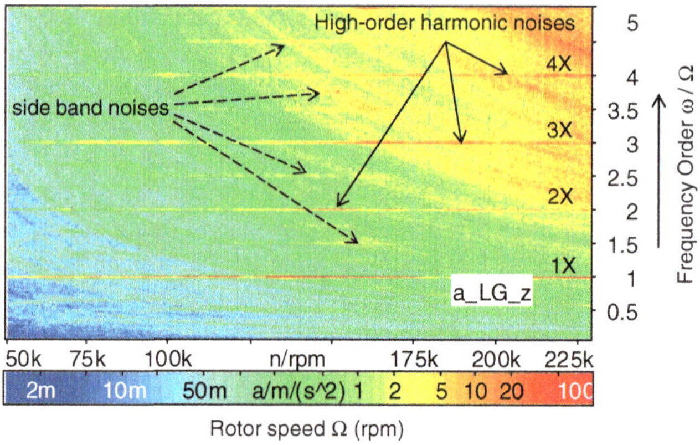

Fig. 4.19 Noise spectra of a ball bearing turbocharger (*Courtesy* BMTS)

However, high-order harmonic noises are induced at large damper oil film thickness. Therefore, the oil film damper set-up mechanism in the bearing housing must be thoroughly studied to improve the noise behavior of automotive turbochargers using the rolling element bearings.

4.5 Wear Noise

Wear defects in the rolling-element bearings, such as defects in the inner and outer races, cage (retainer), and rolling elements (balls or rollers) generate different noise types that have asynchronous high-frequency orders and as well as sidebands frequencies due to frequency modulations. These wear noise frequencies are called the bearing defect-related frequencies [4].

Figure 4.20 shows the key elements and geometries of a ball bearing. The balls with diameter d are held in the cage (retainer); the pitch diameter D_p is defined by the maximum diametral distance between the centers of two opposite balls. Contact angle θ is the angle between the line perpendicular to the shaft and the ball radius at the contact point in the outer race. The inner race is fixed to the shaft and rotates at the rotor speed N; the outer race is mounted in the cartridge supported by the damper oil film.

The bearing defect-related frequencies are computed from the bearing geometries, number of balls, and rotor speed [2, 7].

Fundamental train frequency (*ftf*) is caused by the cage defects depending on the rotor speed N in rps.

$$ftf = \frac{N}{2}\left(1 - \frac{d}{D_p}\cos\theta\right) \tag{4.36}$$

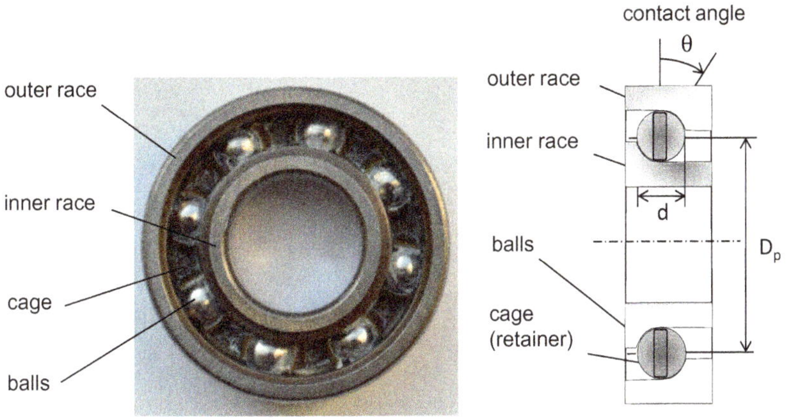

Fig. 4.20 Key elements of an angular contact *ball* bearing [4]

Ball passing frequency over defective inner race (*bpfi*) is generated when the *Z* balls or rollers pass over the defective inner race.

$$bpfi = Z\frac{N}{2}\left(1 + \frac{d}{D_p}\cos\theta\right) \tag{4.37}$$

Ball passing frequency over defective outer race (*bpfo*) is generated when the *Z* balls or rollers pass over the defective outer race.

$$bpfo = Z\frac{N}{2}\left(1 - \frac{d}{D_p}\cos\theta\right) \tag{4.38}$$

Two-times ball spin frequency (2*bsf*) is induced by the defective balls or rollers spinning over the inner and outer races.

$$2bsf = N\frac{D_p}{d}\left[1 - \left(\frac{d}{D_p}\right)^2\cos^2\theta\right] \tag{4.39}$$

where

d is the diameter of balls or rollers;
D_p is the pitch diameter of the bearing;
θ is the contact angle of the rolling elements;
Z is the number of rolling elements;
N is the rotor speed in rps.

The bearing defect-related frequencies are displayed in Fig. 4.21 for a ball bearing containing eight balls with $d = 3.2$ mm, $D_p = 12.2$ mm, and $\theta = 45°$.

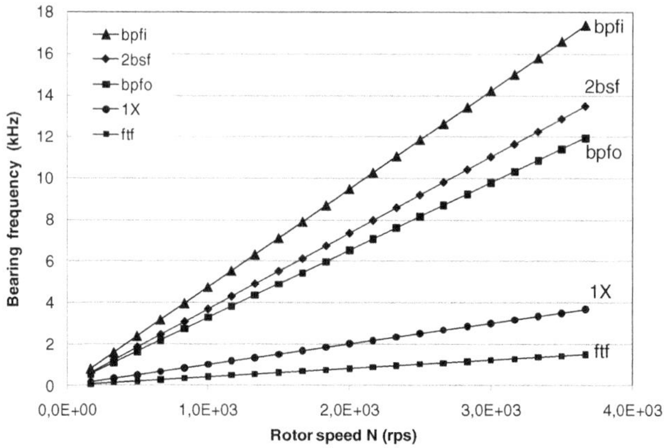

Fig. 4.21 Bearing defect-related frequencies [4]

The frequencies induced bearing defects *bpfi*, *2bsf*, and *bpfo* are supersynchronous (f > 1X); the *ftf* frequency is subsynchronous (f < 1X).

References

1. Cumpsty, N.A.: Compressor aerodynamics. Krieger Publishing Company, Malabar (2004)
2. Ehrich, F.: Handbook of rotordynamics. Krieger Publishing Company, Malabar (2004)
3. Greitzer, E.M.: The stability of pumping systems—the 1980 Freeman scholar lecture. J. Fluids. Eng. **103**, 193–242 (1981)
4. Nguyen-Schäfer, H.: Rotordynamics of automotive turbochargers. Springer, Berlin-Heidelberg (2012)
5. Rieger, N.F.: Rotordynamics 2—problems in turbomachinery. CISM courses and lectures no 297. Springer, Wien, New York (1988)
6. Rieger, N.F.: Balancing of rigid and flexible rotors. U.S. DoD, Washington (1986)
7. Taylor, J., Kirkland, D.W.: The bearing analysis handbook. Vibration Consultants Inc., Tampa (2004)
8. Whitfield, A., Baines, N.C.: Design of radial turbomachines. Pearson Education, Longman Scientific and Technical, Harlow (1990)

Chapter 5
Computational Nonlinear Rotordynamics of Turbochargers

5.1 Vibration Equations of Automotive Turbocharger Rotors

Figure 5.1 displays the rotor of an automotive turbocharger in the center housing and rotating assembly (CHRA) including the rotor shaft, compressor and turbine wheels, rotating floating ring bearings (RFRBs), seal rings, and thrust rings. All rotor components must be taken into account in the rotor dynamic computation to study the rotor vibration response, such as the frequency components in the Waterfall plot, rotor orbits in the phase plane, and response amplitudes in the time domain diagram.

The continuum rotor vibration equations are generally discretized by two schemes: the finite element and transfer matrix methods. The finite element method (FEM), a discretization method of the continuum rotor structure, is based on the principle of D' Alembert, where the sum of virtual work of all rotor elements equals zero. By discretization of the continuum rotor vibration equations, one divides the continuum rotor into a finite number of elements of the computational model that includes the concentrated masses, cylindrical elements, disks, and interfaces between the rotor, bearings, and seal rings. Therefore, the rotor vibration equations are derived and written in the discrete matrix Eq. (5.1). Another alternative method is the transfer matrix method (TMM) developed by Myklestad and Prohl, in which the rotor first element is connected with the last one by the overall transfer matrix that consists of the transfer matrices of all involved elements between the first and last rotor elements. The main advantage of the TMM is to spare the working memory compared to the FEM.

The rotor vibration equations with N degrees of freedom (DOF), as shown in Fig. 5.1 are written in the discrete matrix equation.

$$\mathbf{M}\ddot{\mathbf{x}} + \mathbf{C}_{SG}\dot{\mathbf{x}} + \mathbf{K}_S\mathbf{x} = \mathbf{f}(t) \tag{5.1}$$

H. Nguyen-Schäfer, *Aero and Vibroacoustics of Automotive Turbochargers*,
DOI: 10.1007/978-3-642-35070-2_5, © Springer-Verlag Berlin Heidelberg 2013

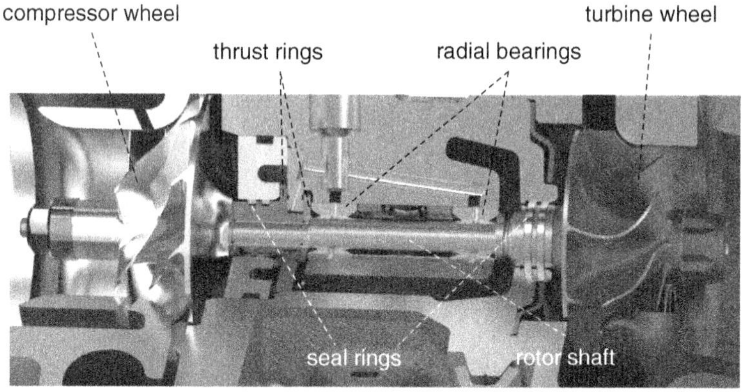

Fig. 5.1 Automotive turbocharger rotor setup in the CHRA (Courtesy BMTS)

where

$\mathbf{M} = (N \times N)$ is the mass matrix containing the rotor masses and inertia moments;
$\mathbf{C_{SG}} = (N \times N)$ is the rotor damping coefficient and gyroscopic matrix;
$\mathbf{K_S} = (N \times N)$ is the shaft stiffness coefficient matrix containing the diagonal and cross-coupled stiffness coefficients;
$\mathbf{x} = (N \times 1)$ is the vibration rotor response vector including all degrees of freedom of two translational and two rotational displacements at each station of the element;
$\mathbf{f}(t) = (N \times 1)$ is the vector of the unbalance forces and moments, and nonlinear bearing forces.

To reduce the second-order to first-order equation system, the $(2N \times 1)$ vector $\mathbf{z}$ of the vibration response is defined by

$$\mathbf{z} \equiv \begin{bmatrix} \mathbf{x} \\ \mathbf{y} \end{bmatrix} = \begin{bmatrix} \mathbf{x} \\ \dot{\mathbf{x}} \end{bmatrix} \Rightarrow \dot{\mathbf{z}} = \begin{bmatrix} \mathbf{y} \\ \dot{\mathbf{y}} \end{bmatrix} = \begin{bmatrix} \dot{\mathbf{x}} \\ \ddot{\mathbf{x}} \end{bmatrix} \tag{5.2}$$

where

$$\mathbf{y} \equiv \dot{\mathbf{x}};\ \dot{\mathbf{y}} \equiv \ddot{\mathbf{x}}.$$

By substituting the vector $\mathbf{z}(t)$ into Eq. (5.1), $2N$ first-order vibration equations of $\mathbf{z}$ result in

$$\begin{aligned} \begin{bmatrix} \dot{\mathbf{x}} \\ \ddot{\mathbf{x}} \end{bmatrix} &= \begin{bmatrix} \mathbf{0} & \mathbf{I} \\ -\mathbf{M}^{-1}\mathbf{K_S} & -\mathbf{M}^{-1}\mathbf{C_{SG}} \end{bmatrix} \cdot \begin{bmatrix} \mathbf{x} \\ \dot{\mathbf{x}} \end{bmatrix} + \begin{bmatrix} \mathbf{0} \\ \mathbf{M}^{-1}[\mathbf{U}(\Omega,t) + \mathbf{F_B}(\mathbf{x},\dot{\mathbf{x}},t)] \end{bmatrix} \\ \Leftrightarrow \dot{\mathbf{z}} &= \begin{bmatrix} \mathbf{0} & \mathbf{I} \\ -\mathbf{M}^{-1}\mathbf{K_S} & -\mathbf{M}^{-1}\mathbf{C_{SG}} \end{bmatrix} . \mathbf{z} + \begin{bmatrix} \mathbf{0} \\ \mathbf{M}^{-1}[\mathbf{U}(\Omega,t) + \mathbf{F_B}(\mathbf{x},\dot{\mathbf{x}},t)] \end{bmatrix} \end{aligned} \tag{5.3}$$

where

$\mathbf{U}$ is the unbalance force and moment;
$\mathbf{F_B}$ is the nonlinear bearing force computed by means of the impedance method that is resulted from solving the two-phase Reynolds lubrication equation, as given in Eq. (5.10).

The coupled nonlinear vibration equation system (5.3) is written in the matrix equation of $\mathbf{z}$.

$$\dot{\mathbf{z}} = \mathbf{A}\mathbf{z} + \mathbf{b}(\mathbf{z},\ \Omega,\ t) \tag{5.4a}$$

where

$\mathbf{A} = (2N \times 2N)$ rotor mass, mass inertia moments, stiffness, and gyroscopic effect matrix;
$\mathbf{b} = (2N \times 1)$ external force vector of unbalance and nonlinear bearing forces. Within

$$\mathbf{A} = \begin{bmatrix} \mathbf{0} & \mathbf{I} \\ -\mathbf{M}^{-1}\mathbf{K_S} & -\mathbf{M}^{-1}\mathbf{C_{SG}} \end{bmatrix}; \tag{5.4b}$$

$$\mathbf{b}(\mathbf{z},\ \Omega,\ t) = \begin{bmatrix} \mathbf{0} \\ \mathbf{M}^{-1}[\mathbf{U}(\Omega) + \mathbf{F_B}(\mathbf{z}, t)] \end{bmatrix} \tag{5.4c}$$

The matrix $\mathbf{A}$ containing the matrices $\mathbf{M}$, $\mathbf{K_S}$, and $\mathbf{C_{SG}}$ is obviously non-symmetric; i.e., $\mathbf{A} \neq \mathbf{A}^\mathrm{T}$ where $\mathbf{A}^\mathrm{T}$ is the transposed matrix of $\mathbf{A}$. For this reason, the bimodal method [1] is used to solve the first-order nonlinear matrix equation of $\mathbf{z}$ with a $(2N \times 2\ N)$ matrix $\mathbf{A}$ of N degrees of freedom.

5.2 Two-Phase Reynolds Lubrication Equation

Figure 5.2 demonstrates a two-phase oil model in the bearing clearances in which the oil mixture volume V_{mix} consists of the volumes of gas bubbles V_B and pure oil liquid V_{liq}. Obviously, one has $V_{\text{mix}} = V_B + V_{\text{liq}}$.

The dynamic viscosity of the oil mixture is calculated from the oil liquid viscosity and gas bubble fraction according to [2].

$$\eta_{\text{mix}} = \eta_{\text{liq}}\left(\frac{1}{1+r}\right) = \eta_{\text{liq}}\theta \tag{5.5}$$

where $r \equiv \frac{V_B}{V_{\text{liq}}}$ is the bubble fraction in the oil liquid.

The bearing filling grade θ is defined by the ratio of the oil liquid to oil mixture volumes; it can be written in the bubble fraction r.

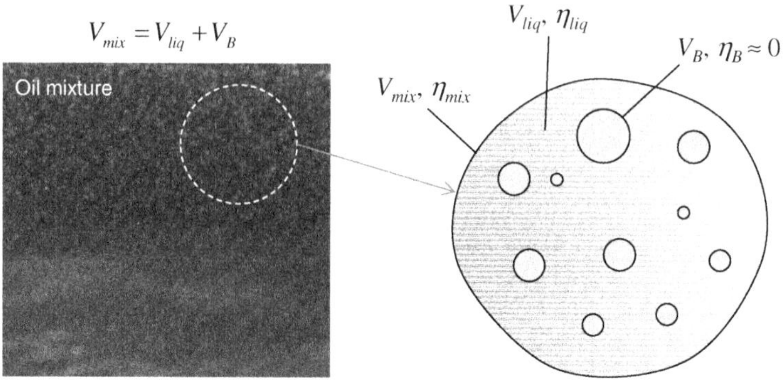

Fig. 5.2 A two-phase oil mixture model

$$\theta \equiv \frac{V_{\text{liq}}}{V_{\text{mix}}} = \frac{V_{\text{liq}}}{V_{\text{liq}} + V_B} = \frac{1}{1+r} \tag{5.6}$$

Similarly, the oil mixture density is calculated from the oil liquid density and bearing filling grade.

$$\begin{aligned} \rho_{\text{mix}} = \frac{m_{\text{mix}}}{V_{\text{mix}}} \approx \frac{m_{\text{liq}}}{V_{\text{liq}} + V_B} &= \frac{m_{\text{liq}}}{V_{\text{liq}}(1 + V_B/V_{\text{liq}})} \\ &= \rho_{\text{liq}} \left(\frac{1}{1+r} \right) = \rho_{\text{liq}} \theta \end{aligned} \tag{5.7}$$

The dynamic viscosity and density of the oil mixture given in Eqs. (5.5 and 5.7) depend only on the gas bubble fraction and are not affected by the dissolved gas in the oil liquid.

There are two cases of the bearing filling grade: first, $\theta = 1$ for the pure oil liquid without air-releasing or cavitation; second, $\theta = 0$ for the oil condition at 100 % gas phase due to air release or cavitation in oil. However, the bearing filling grade lies between 0 and 1 ($0 < \theta \leq 1$) in the most application cases.

Having used the Henry-Dalton's and Boyle-Mariotte's laws, the gas bubble fraction is calculated at T and p by

$$\begin{aligned} r \equiv \frac{V_B}{V_{liq}} &= r_0 \left(\frac{T p_0}{T_0 p} \right) - a_v \left(1 - \frac{T p_0}{T_0 p} \right) \\ &= (r_0 + a_v) \left(\frac{T p_0}{T_0 p} \right) - a_v \end{aligned} \tag{5.8}$$

where

a_v is the Bunsen's coefficient chosen between 0.08 and 0.09 for the automotive mineral oils of ISO-VG 32–220 to 30 MPa. This Bunsen solubility coefficient is defined as the maximum dissolved gas volume per unit oil liquid volume in the equilibrium condition at the gas partial pressure of 101.3 kPa;

r_0 is the initial gas bubble fraction in the oil liquid at the bearing inlet;
T and p are the oil temperature and pressure in the oil film in the operating condition;
T_0 and p_0 are the ambient or oil inlet temperature and pressure.

Substituting Eq. (5.8) into Eq. (5.6), one obtains the bearing filling grade

$$0 < \theta \equiv \frac{V_{liq}}{V_{mix}} = \frac{1}{(1-a_v) + (r_0 + a_v)\left(\frac{Tp_0}{T_0 p}\right)} \leq 1 \tag{5.9}$$

The Reynolds lubrication equation of two-phase flow lubricating oil is written in

$$\frac{\partial}{\partial x}\left(\frac{\rho_{\text{liq}}}{\eta_{\text{liq}}} h^3 \frac{\partial p}{\partial x}\right) + \frac{\partial}{\partial z}\left(\frac{\rho_{\text{liq}}}{\eta_{\text{liq}}} h^3 \frac{\partial p}{\partial z}\right) = 6\left[(U_r + U_b)\frac{\partial}{\partial x}(\rho_{\text{liq}}\theta h) + 2\frac{\partial}{\partial t}(\rho_{\text{liq}}\theta h)\right] \tag{5.10}$$

where

$p(x, z)$ is the current oil film pressure in the bearing clearance;
$\eta_{\text{liq}}(T)$ is the oil liquid viscosity computed from the Cameron and Vogel's equation [4];
U_r and U_b are the circumferential velocities of the rotor and bearing ring;
$h(x)$ is the current oil-film thickness in the bearing clearance;
θ is the bearing filling grade, as given in Eq. (5.9).

5.3 Nonlinear Bearing Forces

The bearing forces F_r and F_t are based on the rotating coordinate system (r, t) fixed to the journal (i.e., shaft inside the bearing). In fact, the bearing forces acting upon the journal in the inertial coordinate system (X_1, X_2) are required for the rotor

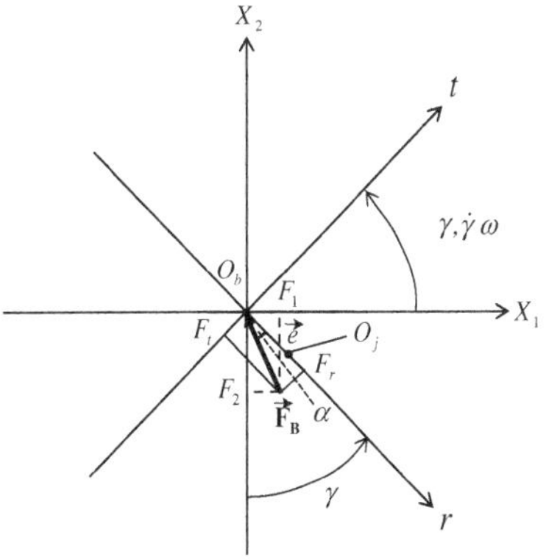

Fig. 5.3 Coordinate transformation of the bearing force components

dynamic vibration equations. Therefore, the transformation from the rotating coordinates into the inertial coordinates is necessary to compute the bearing forces in the inertial coordinate system (X_1, X_2), as given in Fig. 5.3.

The radial and tangential components of the bearing force $\mathbf{F_B}$ acting on the journal, opposite to the journal force $\mathbf{F_J}$, are resulted from integrating the oil pressure $p(\xi, z)$ of Eq. (5.10) on the journal surface.

$$\mathbf{F_B} = \begin{bmatrix} F_r = \int_0^L R \int_0^{2\pi} p(\xi,\ z) \cos \xi \ \mathrm{d}\xi.\mathrm{d}z \\ F_t = \int_0^L R \int_0^{2\pi} p(\xi,\ z) \sin \xi \ \mathrm{d}\xi.\mathrm{d}z \end{bmatrix} = -\mathbf{F_J} \tag{5.11}$$

within

L is the inner bearing width;
R is the journal or shaft radius;
ξ is the position angle for the oil-film thickness where $x = R\xi$(see Fig. 5.4).

The nonlinear bearing forces in Eq. (5.11) are transformed into the inertial coordinate system (X_1, X_2):

$$\mathbf{F_B} = \begin{pmatrix} F_1 \\ F_2 \end{pmatrix} = \begin{pmatrix} \sin\gamma & \cos\gamma \\ -\cos\gamma & \sin\gamma \end{pmatrix} \begin{pmatrix} F_r \\ F_t \end{pmatrix} \tag{5.12}$$

where

F_1, F_2 are the force components in the inertial coordinates X_1 and X_2;
F_r, F_t are the force components in the rotating coordinates r and t.

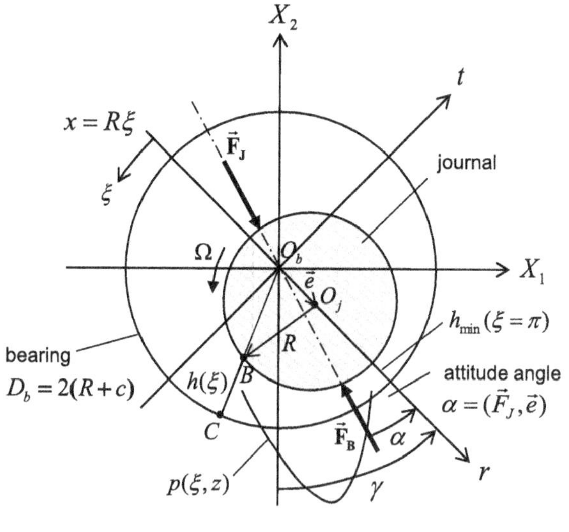

Fig. 5.4 Bearing force characteristics

The nonlinear bearing force components are written in the inertial coordinates:

$$\begin{cases} F_1 = +F_r \sin\gamma + F_t \cos\gamma \equiv f_1(\varepsilon, \dot{\varepsilon}, \gamma, \dot{\gamma}, \Omega) \\ F_2 = -F_r \cos\gamma + F_t \sin\gamma \equiv f_2(\varepsilon, \dot{\varepsilon}, \gamma, \dot{\gamma}, \Omega) \end{cases} \tag{5.13}$$

where

γ is the angular position of the journal;
ε is the relative eccentricity of the journal ($\varepsilon = e/c$);
c is the bearing clearance;
α is the attitude angle;
Ω is the rotor speed.

5.4 Boundary Conditions of Nonlinear Rotordynamics

In the following section, the computation of nonlinear rotordynamics for a typical automotive turbocharger using two separate rotating floating ring bearings has been carried out in the study case. The following characteristics are taken into account running the rotordynamic computational program Madyn 2000 [5], which is based on the extensionally developed module for nonlinear rotordynamics of high-speed turbochargers [4]:

- run-up simulation to very high rotor speeds;
- transient behavior of the rotor responses;
- gyroscopic effect of the flexible rotor;
- two-phase flows of the oil films in the bearing clearances [2, 3, 6];
- unbalance forces and moments acting upon the rotor;
- damping effects of the seal rings;
- nonlinear bearing forces of the rotating floating ring bearings;
- synchronous and asynchronous vibration responses;
- rotating floating ring bearings with changing ring speed ratios;
- computing the ring speed ratios of the bearings;
- induced oil temperatures in the bearing clearances;
- heat convection in the oil films in the axial direction;
- heat conduction between the oil-films in the radial direction;
- various types of lubricating oil (SAE 0W30, 5W30, 10W40, 15W40);
- different pressures and temperatures of lubricating oil.

The computed rotor of an automotive turbocharger for passenger vehicles consists of the compressor and turbine wheels, rotor shaft, thrust rings, seal rings, and rotating floating ring bearings. One transforms this computed rotor into the discretized finite element model where the components of the rotor, such as the compressor and turbine wheels, rotor shaft, and radial bearings are discretized in a finite number of the concentrated mass points, disks, cylinder elements, and interfaces of the bearings and seal rings (see Fig. 5.5). The finite element model of the computed

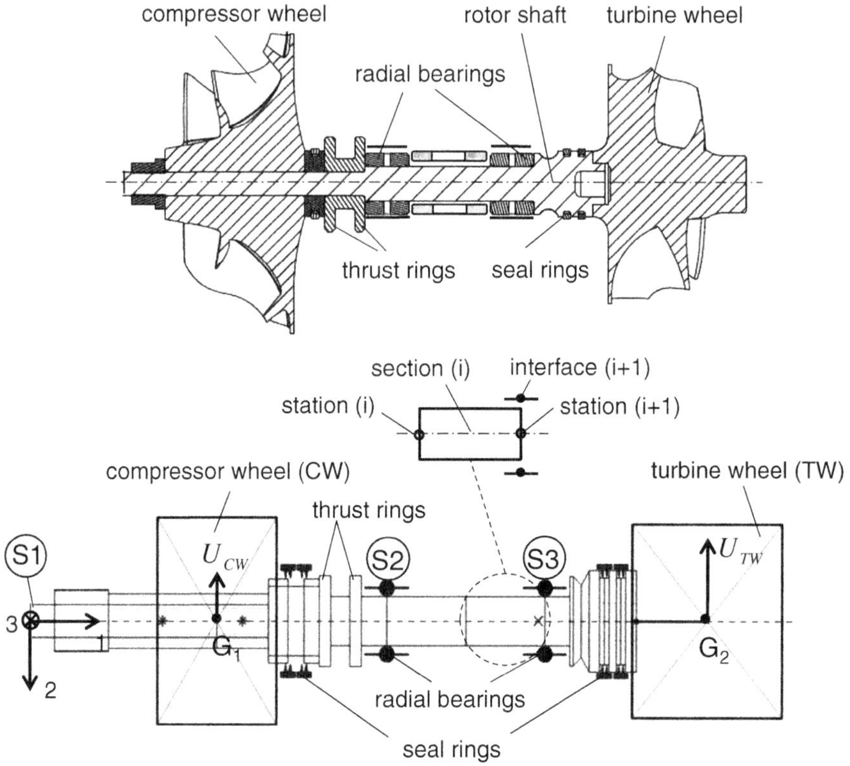

Fig. 5.5 Computing rotor model and its finite element model [4]

rotor contains many sections that are connected by the stations at both ends of each section. By solving the two-phase Reynolds lubrication equation of two-oil-film bearings, the impedance table for the bearing forces is created for the bearing force computation. The nonlinear bearing forces in the bearing clearance are resulted from the corresponding bearing stiffness and damping coefficients by linearly interpolating the computed values of the bearing characteristics at the journal center position (ε,γ) and Sommerfeld number *So* in the impedance table. The nonlinear bearing forces contain two force components: the rotation force induced by the rotor speed Ω depends on the bearing force angle α and *So* (η,ε,γ); the damping force caused by the journal velocities $\dot{\varepsilon}, \dot{\gamma}$ depends on $\eta, \varepsilon, \gamma, \dot{\varepsilon}, \dot{\gamma}$, as given in [4].

5.5 Computational Results of Turbochargers

The nonlinear bearing forces acting upon the journal are computed at each iteration step based on the Sommerfeld number *So* by the impedance method. The coupled nonlinear vibration equations of the rotor are given in Eq. (5.3), in which we put the nonlinear bearing forces, as given in Eqs. (5.11 and 5.13) on the right-hand side of Eq. (5.3). By using the bimodal method [1], we decouple the coupled nonlinear equations in the decoupled nonlinear equations. Then, the vibration responses are solved by a high-order Runge–Kutta scheme at each iteration step.

Using the Discrete Fourier Transform (DFT) with FFT, the frequency spectrum of the rotor responses is displayed in the spectrogram (Waterfall plot). The rotor orbits at the station S1 of the compressor inlet and at the stations S2 and S3 of the journal loci in the radial bearings are computed in the runup simulation, as shown in Fig. 5.5.

Two rotating floating ring bearings and two seal rings at each side support the computed rotor with a mass at about 150 g including the shaft with a diameter of 7 mm, compressor wheel, turbine shaft, and thrust rings. The relative inner diametral clearance of the bearing is initially about 3×10^{-3} due to the temperature difference between the inner and outer oil-films, the centrifugal force of the bearing ring, and the bearing ring expansion in the radial direction. Hence, the relative inner diametral clearance increases to a maximum about 4×10^{-3} during operation. The ratio of the outer and inner bearing clearances is varied by a factor between three and four. The maximum rotor speed is limited to nearly 225,000 rpm ($\Omega = 3{,}750$ Hz).

The compressor wheel and turbine shaft have residual unbalances of approximately 0.1 and 0.2 g.mm, respectively. The lubricating oil SAE 5W30 is used in the runup computation; the oil dynamic viscosity changes with the effective oil-film temperatures in every iteration step according to the Cameron and Vogel's equation [4]. The oil condition at the inlet is given at a relative pressure of 0.3 MPa and oil temperature of 90 °C at S2; 100 °C at S3.

The computing time for the run-up simulation of about 20-hour-CPU is necessary on a PC with a dual-core AMD Opteron of 3 GHz with 8 GB RAM. This computing time is quite acceptable in the industry instead of a-few-week CPU in the case of instantaneously solving the Reynolds equation at each iteration step.

The vibration mode of the flexible rotor in the run-up simulation is analyzed at the stations S2 and S3. Figure 5.6 displays the vibration modes of the rotor at various rotor speeds. At low rotor speeds, the rotor vibration has a conical mode. From the rotor speed of about 81,000 rpm (36 % relative rotor speed) – 135,000 rpm (60 % relative rotor speed), the rotor vibration changes from the conical to the cylindrical mode.

At low rotor speeds up to nearly 45,000 rpm, the rotor orbits at S2 and S3 are mostly eccentric to each other (see Fig. 5.7), i.e., the mode shape of the rotor response is conical since the rotor is still rigid at low rotor speeds. At high rotor speeds from about 81,000 rpm, the bearing stiffness and damping coefficients increase, the conical mode of the rotor vibration tends toward cylindrical mode. Figure 5.7 shows that orbits at S2 and S3 are nearly concentric from the relative speed of 60 % up to the maximum rotor speed. The limit cycles of the rotor response at S2 and S3 have peak-to-peak amplitude of about 25 μm at 135,000 rpm and reduce to 15 μm at the maximum rotor speed of 225,000 rpm.

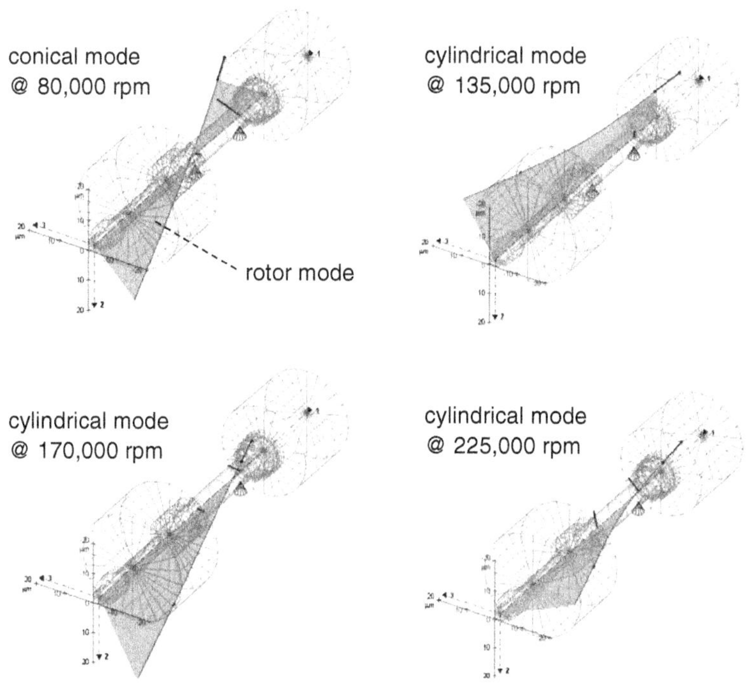

Fig. 5.6 Vibration modes of the turbocharger rotor [4]

The unbalance amplitude (1X) is relatively small compared to the asynchronous amplitudes of the inner and outer oil whirls in the entire rotor speed range.

In fact, the typical resonance due to unbalance does not exist in nonlinear rotordynamics (see Figs 5.8, 5.9a, and b), but only the limit cycle of the rotor response occurs at each rotor speed. The reason is the effective rotor stiffness coefficient depends not only on the shaft stiffness coefficient itself but also on the bearing stiffness coefficient. The rotor equivalent stiffness coefficient is called the nonsynchronous complex dynamic stiffness. Therefore, the bearing stiffness coefficient is dependent on the effective oil pressure, temperature, rotor speed, bearing ring speed, and journal eccentricity as well. At large rotor deflections, the journal eccentricity increases in the bearing clearance; the oil-film thickness reduces in the bearing clearance to generate the large bearing force. As a result, the bearing stiffness and damping coefficients increase with the rotor deflection until the journal orbit stabilizes after the Hopf bifurcation point in the limit cycle at each rotor speed [4]. In this case, no typical resonance like in linear rotordynamics occurs in the entire rotor speed range, as shown in the Waterfall plot (see Fig. 5.8).

In practice, one plots Waterfall plot in a two-dimensional plot (modified Waterfall plot), in which the amplitudes of the vibrations are displayed in color or grayscale in the frequency order spectra, as shown in Fig. 5.9a. Instead of the whirl frequency ω, the frequency order ω/Ω versus rotor speed Ω is used in the modified

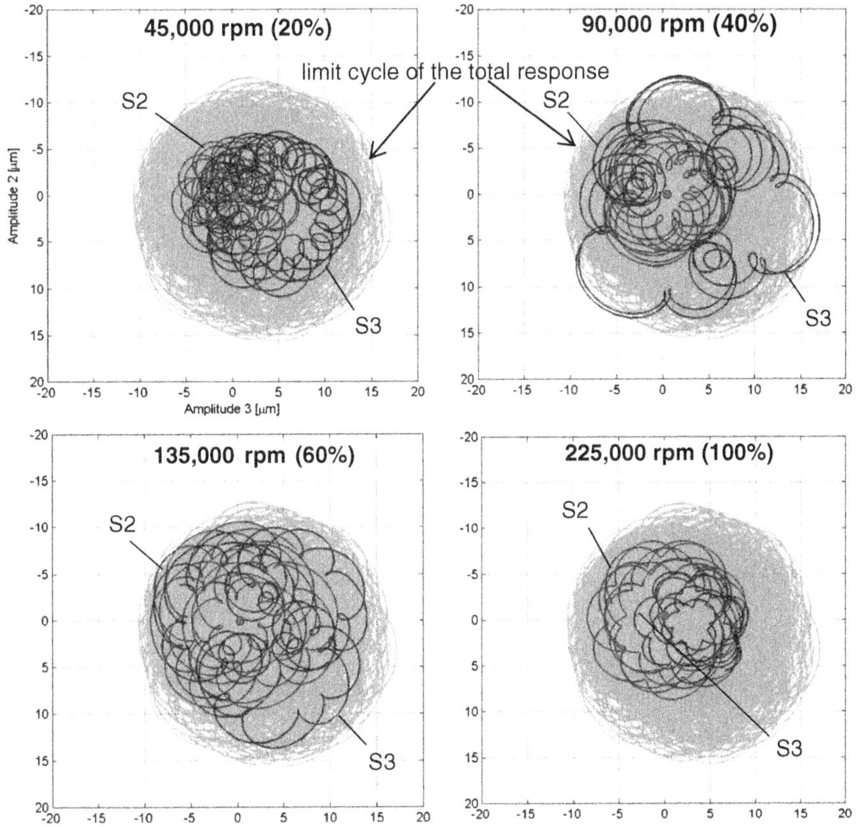

Fig. 5.7 Orbit of the journal at the stations S2 and S3 [4]

Waterfall plot. Its main advantage is it is easier to analyze the frequency order spectra of the synchronous and asynchronous vibrations at the rotor speed. The frequency jump of the inner oil whirl takes place at nearly 30 % of relative rotor speed, where the conical vibration mode changes into the cylindrical mode.

Figure 5.9a and b shows the comparison between the computation of rotordynamics and measurement results in Waterfall plots. Compared to the measurement, the computed inner oil whirl takes place a little later at the relative rotor speed of about 30 % (67,500 rpm) instead of at nearly 27 % (62,000 rpm) as in the measurement. The frequency order of the computed inner oil whirl (inner OW) begins at a lower frequency order of 0.4X compared to about 0.55X in the measurement and ends at the frequency order of 0.2X, nearly the same as in the measurement results. The low frequency order of the inner oil whirl is caused by the higher

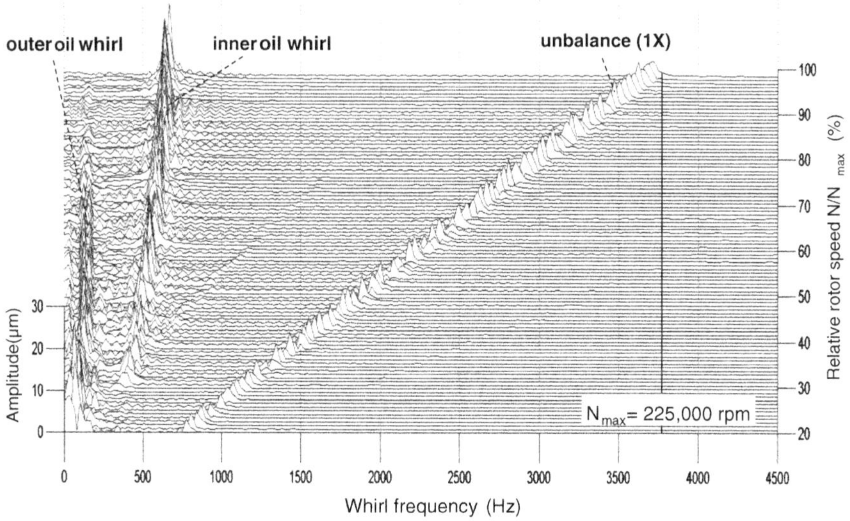

Fig. 5.8 Waterfall plot of the journal at the station S1 [4]

bearing ring speed in the measurement than in the computation. The unknown heat flow transferred from the bearing housing to the outer oil-film of the bearing has not been taken into account in the computation. For this reason, the outer oil-film temperature increases in the measurement compared to the computation; therefore, the ring speed ratio of the bearing is higher, leading to the higher frequency order of the inner oil whirl.

In fact, the residual unbalance vector directions of the compressor and turbine wheels are unknown after the trim balancing in the measurement. For this reason, the resulting residual unbalance may be smaller in the computation. Thus, the frequency jump of the inner oil whirl occurs at a rotor speed of about 67,500 rpm a little later than in the measurement at 62,000 rpm. Similarly, the computed frequency order of the outer oil whirl (outer OW) is a little lower than the frequency order as in the measurement due to the reduced bearing ring speed, especially at low rotor speeds.

In the rotating floating ring bearing case, the journal moves relatively to the bearing ring at every rotor speed; i.e., the journal motion pushes the bearing ring in the radial direction. In fact, it is very difficult to measure the relative displacements between the journal and bearing ring during the rotation. Hence, the rotordynamic computation is a useful tool to determine the relative displacements and the current oil-film thickness in the rotating floating ring bearings at any rotor speed. Figure 5.10 displays the relative displacements between the journal and bearing ring at different rotor speeds.

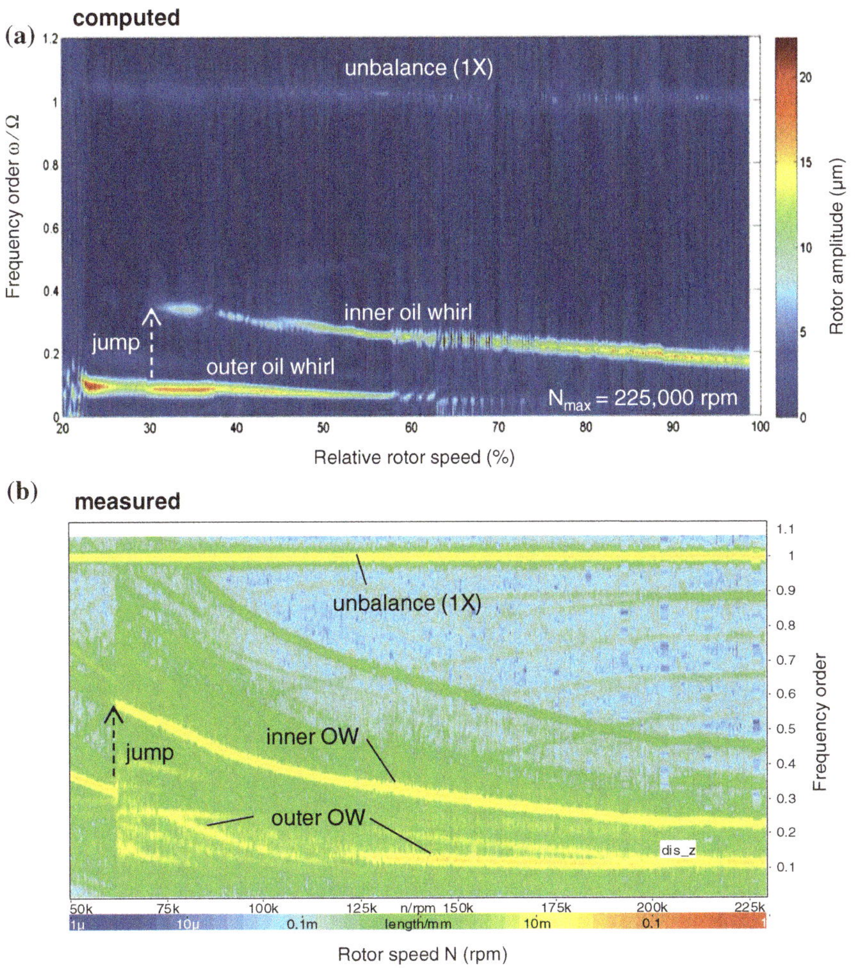

Fig. 5.9 **a** and **b** Waterfall plots of the journal vibration at S1 [4]

At the beginning, the journal and bearing ring are nearly concentric to each other since the rotor mass is very small at nearly 150 g. If the relative displacement is positive, the journal moves closer to the bearing ring because the journal displacement is larger than the ring displacement. In the other case, the journal moves away from the bearing ring. The current oil-film thickness is resulted from the bearing clearance and relative displacement. When the oil-film thickness is larger than the limit oil-film thickness [4], the lubrication regime is fully hydrodynamic and outside the mixed and boundary lubrications; little or no wear occurs in the bearing. The computed results show that the maximum relative displacement is about 9 μm in the limit cycle for the entire rotor speed range. It indicates that the minimum oil-film thickness in the inner bearing clearance is nearly 5 μm with the

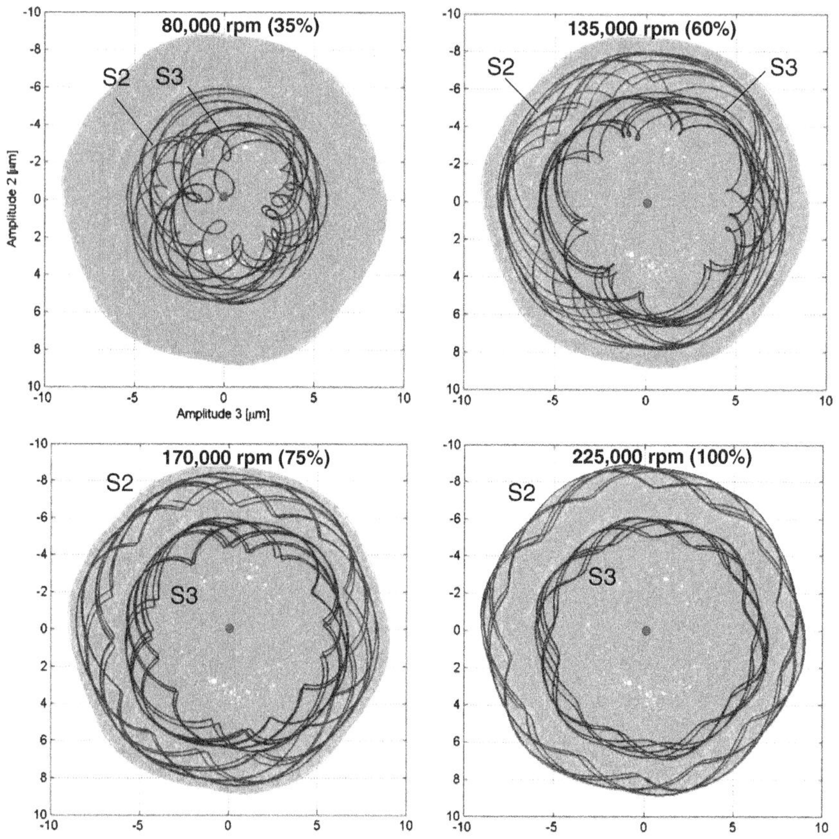

Fig. 5.10 Waterfall plots of the relative displacements between the journal and bearing ring at S2 and S3 [4]

inner radial clearance of 14 μm. The minimum oil-film thickness of 5 μm in the bearing clearance is larger than the limit oil-film thickness; hence, no wear occurs in the bearing.

References

1. Nordmann, R.: Schwingungsberechnung von nichtkonservativen rotoren mit hilfe von links- und rechts-eigenvektoren (in German), VDI Bericht 269 (1976)
2. Bartel, D.: Simulation von tribosystem (in German). Springer-Teubner, (2010)
3. Nguyen-Schäfer, H.: Nonlinear rotordynamics computation of automotive turbochargers using rotating floating ring bearings at high rotor speeds. 10th SIRM International Conference, Berlin (2013)
4. Nguyen-Schäfer, H.: Rotordynamics of automotive turbochargers. Springer, Berlin-Heidelberg (2012)

5. Schmied, J: Program Madyn 2000 for computation of rotordynamics, Delta J.S., Zurich (2011)
6. Nguyen-Schäfer, H., Sprafke, P.: Numerical study on interaction effects of the bubbles induced by air-release and cavitation in hydraulic systems, tenth bath international fluid power workshop bath. UK Research Studies Press Ltd, England (1997)

Chapter 6
Subsynchronous Constant Tone

Constant tone is induced by the inner oil whirl in hydrodynamic radial bearings. The inner oil whirl frequency order in the rotating floating ring bearings reduces from 0.4X to about 0.3X at increasing rotor speeds, at which the inner oil film temperature increases, leading to reduced oil viscosity. Therefore, the inner oil whirl frequency varies in a narrow frequency band from 600 to 1,000 Hz of the human audible range. This frequency is usually considered in automotive turbochargers as *quasi-constant* compared to the rotor frequency; hence, its noise is called *constant tone*. The constant tone occurs at the engine speed range between 1,500 and 3,500 rpm in second to fifth gears with middle to high loads.

Figure 6.1 shows the unbalance whistle and constant tone acceleration frequency spectra in the Waterfall plot. Oil whirl is the self-excited instability with a subsynchronous frequency that usually occurs in automotive turbochargers. The oil whirling in the bearing clearance causes the oil whirl; therefore, it has nothing to do with the rotor unbalance. This means that the oil whirl happens in the bearing without precondition whether the rotor is well balanced or unbalanced. However, a strongly unbalanced rotor could reduce the oil whirl amplitude where the rotor is close to the bearing wall.

The constant tone level is measured by an acceleration sensor at the bearing center housing, as displayed in Fig. 6.1. The constant tone is a subsynchronous vibration response; i.e., its whirl frequency is lower than the rotor frequency. For this reason, it is a subsynchronous noise type. As long as the damping force in the bearing is larger than the destabilizing force caused by the bearing cross-coupled stiffness coefficient, the journal orbit is in a quasi-stable condition. Hence, the constant tone level is inaudible in the vehicle cabin. The journal begins whirling in the bearing when the destabilizing force exceeds the damping force at increasing rotor speeds, leading to the self-excitation instability of the subsynchronous inner oil whirl, which generates the constant tone. Note that the noise intensity is proportional to the whirl frequency squared. The outer oil whirl frequency is much lower than the inner oil whirl frequency; therefore, the outer oil whirl noise intensity is negligibly small. As a result, the outer oil whirl noise is inaudible although its whirl amplitude is much larger than the inner oil whirl amplitude.

H. Nguyen-Schäfer, *Aero and Vibroacoustics of Automotive Turbochargers*,
DOI: 10.1007/978-3-642-35070-2_6, © Springer-Verlag Berlin Heidelberg 2013

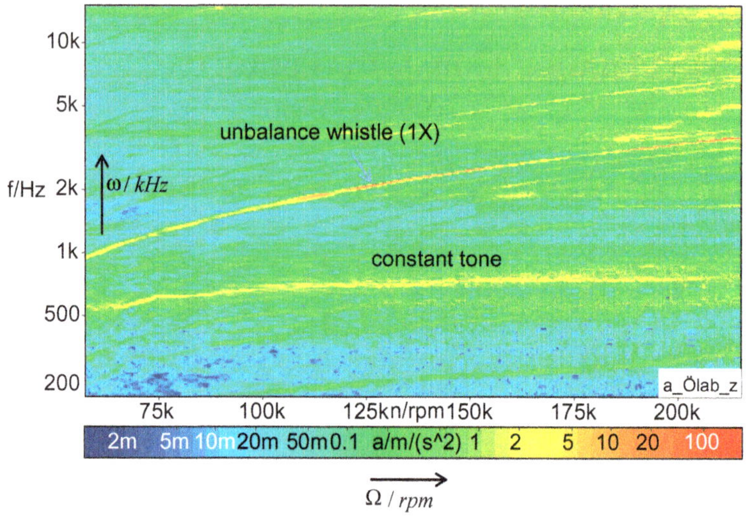

Fig. 6.1 Acceleration frequency spectra in Waterfall plot measured at the bearing center housing [4]

6.1 Transmitting Way of Airborne Noise

Synchronous and asynchronous (super- and subsynchronous) noises induced in automotive turbochargers are caused by compressor aerodynamics of compressor and turbine wheels rotation, and rotordynamics of rotor unbalance and inner oil whirl in the radial oil film bearings. Induced noises transmit through the bearing oil films, bearing center housing, compressor housing, air filter, charge-air intercooler, exhaust-gas manifold, exhaust-gas system (catalytic converter, particle diesel filter, and muffler), and car frame to the vehicle cabin, as shown in Fig. 6.2. On the transmitting ways, they excite the periphery components neighboring the turbocharger. When the noise source frequencies equal the component eigenfrequencies, resonances take place with high response amplitudes in low damping cases (see Chap. 7). The vibration responses of the bearing center housing, air filter, charge-air intercooler, and exhaust-gas system additionally emit airborne noises to the car cabin and environment. Airborne noises are uncomfortable and undesirable for driver and passengers; therefore, they must be reduced as much as possible in passenger vehicles.

The airborne noise level individually depends on the vehicle types because their eigenfrequencies, damping and stiffness coefficients of the periphery components, and car frame are non-identical to each other; therefore, these lead to different noise responses. As a result, the vehicle characteristics play an important role in the airborne noise level of automotive turbochargers. Note that some airborne noises of a turbocharger are inaudible in one vehicle type, but they could be audible in another vehicle type. For this reason, the airborne noise behavior of turbochargers must be individually proofed for each vehicle type in their applications.

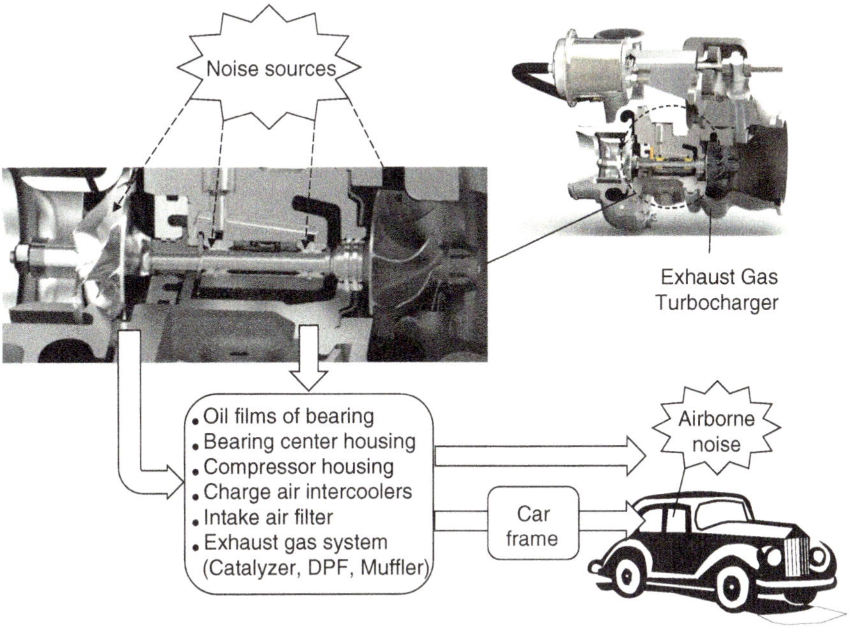

Fig. 6.2 Transmitting way of airborne noise

6.2 Constant Tone Root Causes

The inner oil whirl in the radial bearing is to blame for the constant tone, as earlier discussed. To better understand the constant tone, the inner oil whirl root causes must be analyzed. In the following section, the dynamic force model for the journal and flow model for rotating floating ring bearings are used to study the constant tone. Such analytical models could answer the questions, such as why and when the inner oil whirl takes place in the bearing, which influential parameters the constant tone depend on, and how the constant tone could be reduced in passenger vehicles. Eventually, measures of constant tone reduction are discussed.

At first, the model of dynamic forces acting upon the journal in the rotating coordinate system (r, t) fixed to the journal is displayed in Fig. 6.3. In the radial direction r, the journal stiffness force $-Ke$, bearing cross-coupled damping force $\mathrm{j}c_c\dot{e}$, inertial force $-m\ddot{e}$, and centripetal force $mr_\varepsilon\omega^2 e^{\mathrm{j}\omega t}$ act upon the journal. The destabilizing force $\mathrm{j}k_c e$ caused by the cross-coupled stiffness coefficient k_c of the oil film and the damping force of the bearing $-c\dot{e}$ act upon the journal in the tangential direction t. The inner oil whirl does not occur as long as the damping force is larger than the destabilizing force, as shown in Fig. 6.3. Thus, the stability condition for the inner oil whirl is formulated in the tangential direction:

$$c\dot{e}(t) > \mathrm{j}k_c e(t) \tag{6.1}$$

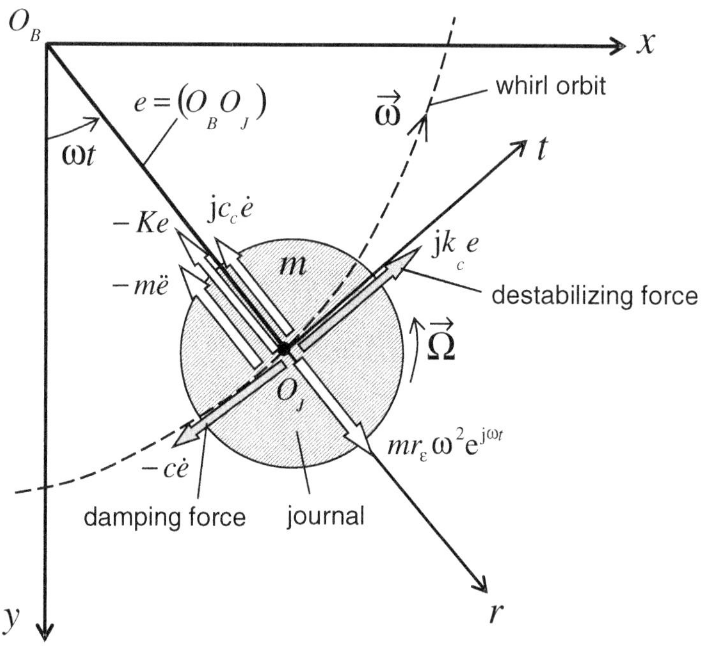

Fig. 6.3 Acting forces on the journal of an oil film bearing

where

$\dot{e} = j\omega e$ is the journal eccentricity velocity;

ω is the whirl speed;

$k_c = c\lambda\Omega$ is the cross-coupled stiffness coefficient of the oil-film bearing within Ω is the rotor speed; λ is defined by the average fluid velocity ratio according to [1]. The average fluid velocity ratio λ versus the journal relative eccentricity ε is displayed in Fig. 6.3.

By using the rotating floating ring bearing with two oil films, two subsynchronous vibration responses of the inner and outer oil whirls induce in the bearing next to the rotor unbalance harmonic vibration response. The constant tone is induced by the inner oil whirl occurring in the radial bearing oil film. At increasing rotor speeds, the journal begins whirling in the inner bearing clearance when the damping force is less than the destabilizing force. In this case, the non-dissipated whirling energy excites the rotor to the self-excited unstable response (inner oil whirl) at a whirl frequency ω.

The stability condition for the inner oil whirl can be derived from Eq. (6.1) and is written in

$$\omega > \lambda\Omega. \tag{6.2}$$

Equation (6.2) indicates that the rotor remains stable as long as the whirl angular velocity ω is larger than $\lambda\Omega$, in which the feeding kinetic energy of the destabilizing force is mostly dissipated by the damping energy in the radial bearing.

At high rotor speeds, the effective temperature of the inner oil film rises due to bearing friction, leading to the oil viscosity decrease in the inner bearing clearance and the reduction of dissipated damping energy as well. As soon as $\omega < \lambda\Omega$, the rotor becomes unstable and begins to whirl in the whirl orbit at the whirl frequency ω (see Fig. 6.3). In the case of using the rotating floating ring bearings, the stability condition for the inner oil whirl results in

$$\begin{aligned} &\omega > \lambda(\Omega + \Omega_R) \\ \Rightarrow\ &\omega > \lambda\Omega\,(1 + RSR) \end{aligned} \tag{6.3}$$

where

RSR $= \Omega_R/\Omega$, is the ring speed ratio of the bearing;

Ω_R is the ring angular speed;

Ω is the angular rotor speed.

In the case of the semi-floating ring or fixed bearings, RSR equals zero; thus Eq. (6.3) becomes the earlier stability condition (q.e.d.) given in Eq. (6.2).

On the one hand, the destabilizing force increases with rotor speed since the bearing cross-coupled stiffness coefficient is proportional to the rotor speed. On the other hand, the damping force in the bearing decreases with rotor speed because the inner oil film temperature in the bearing increases due to bearing fiction; therefore, its viscosity strongly reduces at increasing oil temperatures. As a result, the damping force becomes small and underlies the destabilizing force in the bearing at high rotor speeds. This causes the inner oil whirl, which is a subsynchronous forward whirl; its precession has the same rotor velocity direction. The threshold of instability of the rotor is studied and analyzed in the following subsection.

Having applied Newton's second law to the rotor, the rotor vibration equation is written in the rotating coordinate system (r, t) fixed to the journal.

$$m\ddot{e} + (c - \mathrm{j}c_c)\dot{e} + (K - \mathrm{j}k_c)e = mr_\varepsilon\omega^2 e^{\mathrm{j}\omega t} \tag{6.4}$$

where

m is the rotor mass;

j is the imaginary unit ($\mathrm{j}^2 = -1$);

e is the journal eccentricity;

c is the bearing diagonal damping coefficient;

c_c is the bearing cross-coupled damping coefficient

K is the effective stiffness coefficient of the rotor-bearing system;

$k_c = c\lambda\Omega$ is the bearing cross-coupled stiffness coefficient;

ω is the whirl speed;

r_ε is the rotor unbalance radius.

The rotor response is assumed as an exponential function of the eigenvalue s and time t.

$$e(t) = Ae^{st} \tag{6.5}$$

where

s is the complex eigenvalue including the real part α and the imaginary part ω_n.

$$s = \alpha \pm j\omega_n \tag{6.6}$$

By twice differentiating Eq. (6.5), one achieves

$$\begin{aligned} \dot{e} &= sAe^{st} = se(t); \\ \ddot{e} &= s^2Ae^{st} = s^2e(t) \end{aligned} \tag{6.7}$$

Substituting Eqs. (6.5) and (6.7) into the homogeneous Eq. (6.4), the rotor response characteristic equation is given by

$$D(s) \equiv ms^2 + (c - jc_c)s + (K - jc\lambda\Omega) = 0 \tag{6.8}$$

The rotor stability condition is satisfied if the real part of the complex eigenvalue s is always negative. The real part α is calculated from the characteristic equation (see Appendix C).

$$\alpha = -\frac{c}{2m} \pm \frac{1}{\sqrt{2}}\left[-E + \sqrt{E^2 + F^2}\right]^{\frac{1}{2}} < 0 \tag{6.9}$$

within

$$E \equiv \frac{K}{m} - \left(\frac{c^2 - c_c^2}{4m^2}\right); \quad F \equiv \frac{c}{m}\left(\lambda\Omega - \frac{c_c}{2m}\right) \tag{6.10}$$

By solving Eq. (6.9), one obtains the stability condition for the inner oil whirl.

$$\omega = \frac{c_c}{2m} + \sqrt{\frac{K}{m} + \left(\frac{c_c}{2m}\right)^2} > \lambda\Omega \tag{6.11}$$

At low rotor speeds, the rotor stability condition in Eq. (6.11) is satisfied because the rotor orbit is relatively small and quasi-stable. As the rotor speed increases, the inner oil film temperature rises due to bearing friction, which leads to decrease of the cross-coupled damping and bearing stiffness coefficients. This causes the reduced whirl frequency ω given in Eq. (6.11). Therefore, the rotor stability condition is no longer satisfied because the whirl frequency ω underlies $\lambda\Omega$ at high rotor speeds. In this case, the oil whirl orbit becomes unstable, leading to the self-excited oil whirl instability, which corresponds to the constant tone.

The threshold of instability frequency is given by setting the real part α equal zero, where the oil whirl instability begins.

$$\Omega_{\text{th}} = \frac{1}{\lambda}\left(\frac{c_c}{2m} + \sqrt{\frac{K}{m} + \left(\frac{c_c}{2m}\right)^2}\right) \tag{6.12}$$

Oil whip, a self-excited subsynchronous instability occurs at the rotor speed in which the oil whirl frequency ω equals the first bending critical frequency Ω_{cr}. The oil whip is very dangerous and destructive because the rotor defection continuously exponentially increases with time in the bearing clearance even at a constant rotor speed, leading to the bearing damage due to wear or seizure of the journal in the bearings.

6.3 Constant Tone Analysis

In the following section, influential parameters of the constant tone are studied in the flow model of a two-oil-film radial bearing. The rotor angular speed is defined by Ω; the ring speed of the rotating floating ring bearing by Ω_R. At the journal eccentricity e, the inlet bearing clearance is obviously equal to $(c + e)$; the outlet bearing clearance equals $(c - e)$, as displayed in Fig. 6.4.

Having considered the inner bearing convergent wedge, the oil volumetric flow rate at the inlet is computed.

$$\dot{Q}_{\text{in}} = v_{\text{in}} A_{\text{in}} = \lambda_i [\Omega R + \Omega_R (R + c)](c + e) L_i \tag{6.13}$$

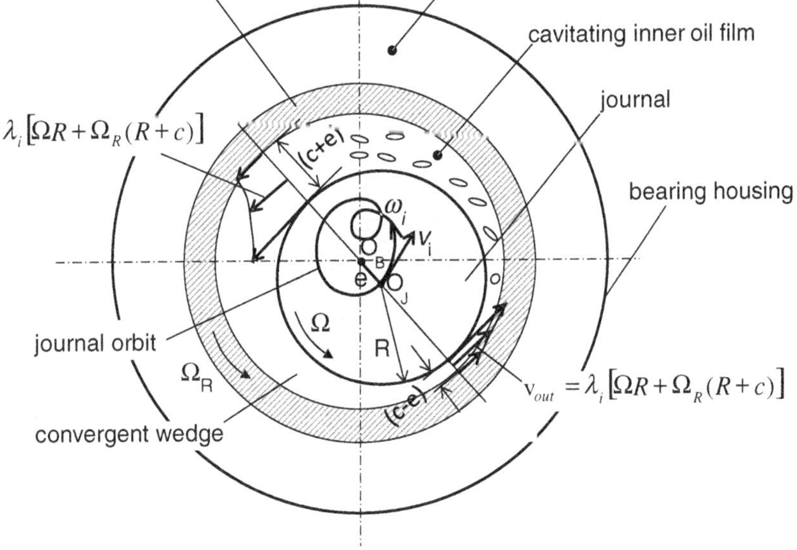

Fig. 6.4 Oil velocity profiles in the bearing clearance

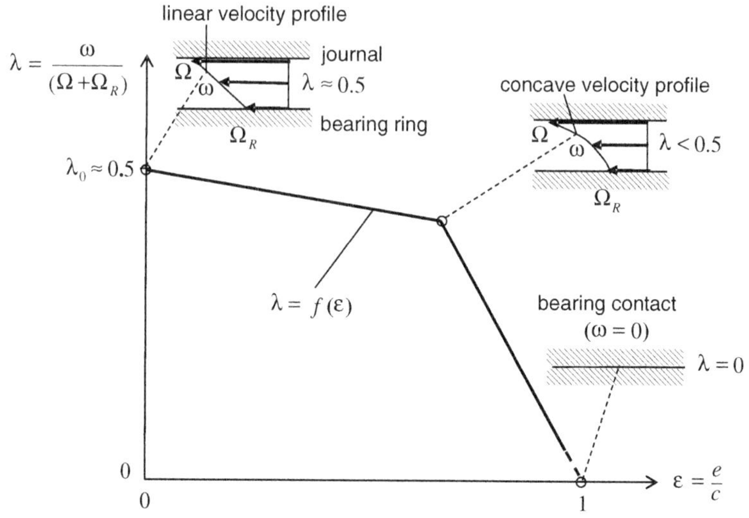

Fig. 6.5 Average fluid velocity ratio λ versus journal relative eccentricity ε

where

λ_i is the average fluid velocity ratio, as given in Fig. 6.5;
c is the radial inner bearing clearance;
e is the journal eccentricity;
L_i is the inner bearing width;
R is the journal radius.

Analogous to Eq. (6.13), the oil volumetric flow rate at the outlet results in

$$\dot{Q}_{\text{out}} = v_{\text{out}} A_{\text{out}} = \lambda_i [\Omega R + \Omega_R (R + c)](c - e) L_i \tag{6.14}$$

According to Eqs. (6.13) and (6.14), the oil inflow rate into the convergent wedge is more than the oil outflow rate. However, the supply oil can only flow in the axial direction toward the bearing ends, but the axial oil flow rate is very small due to the small inner bearing clearance. As a result, the journal has to move at the whirl velocity v_i tangent to the whirl orbit in order to satisfy the mass balance in the inner bearing clearance. The journal whirl frequency ω_i corresponds to v_i, as shown in Fig. 6.4.

The oil surplus volumetric flow rate in the convergent wedge is given by

$$\begin{aligned} \Delta \dot{Q} &= \dot{Q}_{in} - \dot{Q}_{out} \approx 2\mathrm{RL}_i e \lambda_i (\Omega + \Omega_R) \\ &= 2\mathrm{RL}_i e \lambda_i \Omega (1 + \mathrm{RSR}) \end{aligned} \tag{6.15}$$

where $RSR = \Omega_R/\Omega$ is the ring speed ratio of the bearing. The ring speed ratio depends not only on the bearing geometries but also on the effective oil temperatures in the bearing clearances during operation (see Sect. 6.4).

At the journal whirling, the removed oil volumetric flow rate (i.e. oil whirling volumetric flow rate) results in

$$\dot{Q}_w = v_i A_i = (e\omega_i)(2\mathrm{RL}_i) = 2\mathrm{RL}_i e\omega_i \tag{6.16}$$

where

A_i is the displaced journal cross-sectional area; $A_i = 2\mathrm{RL}_i$

ω_i is the journal whirl angular frequency.

Having used the continuity equation in the inner bearing convergent wedge,

$$\sum_i \dot{Q}_i = \Delta\dot{Q} - \dot{Q}_w - \dot{Q}_{ax} = 0 \tag{6.17}$$

The oil whirling volumetric flow rate in Eq. (6.15) is rewritten in

$$\begin{aligned}\dot{Q}_w &= 2\mathrm{RL}_i e\omega_i \\ &= 2\mathrm{RL}_i e\lambda_i \Omega(1+\mathrm{RSR}) - \dot{Q}_{ax}\end{aligned} \tag{6.18}$$

The inner oil whirl amplitude is proportional to the oil whirling volumetric flow rate given in Eq. (6.18).

The journal forward whirl speed is resulted from Eq. (6.18) in

$$\omega_i = \lambda_i \Omega(1+\mathrm{RSR}) - \frac{\dot{Q}_{ax}}{2\mathrm{RL}_i e} \tag{6.19}$$

Since the oil flow rate in the axial direction is negligibly small, the inner oil whirl frequency near the bearing center at $\lambda_i \approx 0.5$ becomes

$$\omega_i = \lambda_i \Omega(1+RSR) \approx \frac{1}{2}\Omega(1+RSR) \tag{6.20}$$

In the case of semi-floating ring or fixed bearings at RSR = 0, the constant tone frequency is nearly one-half rotor frequency ($\omega_i \approx 0.47\Omega$). It is usually called the half-frequency oil whirl. The average fluid velocity ratio of rotating floating ring bearings is qualitatively displayed in Fig. 6.5.

6.4 Calculating Ring Speed Ratio

Floating ring bearings are usually applied to automotive turbochargers to reduce airborne noise and rotor amplitudes at resonances. There are two common types of floating ring bearings: semi-floating (SFRB) and rotating-floating ring bearings (RFRB). The common characteristic of both floating ring bearings is having two oil films: inner and outer oil films. However, the semi-floating bearing ring is non-rotating but floating; the rotating-floating bearing ring is rotating and floating at once.

In order to reduce the bearing friction, the rotating floating ring bearings (RFRB) are used in automotive turbochargers. Due to reduction of the bearing friction at low rotor speeds, the transient response of turbochargers is improved in low-end torque. Figure 6.2 shows the CHRA setup with the rotating floating ring bearings. Lubricating oil at the pressure p_i from the inlet flows through the outer bearing clearance in the circumferential and axial directions and leaves the bearing ends at the ambient pressure p_O. By means of oil holes in the bearing ring, lubricating oil enters the inner bearing clearance, flows in the circumferential and axial directions, and leaves the bearing ends at the ambient pressure p_O. The volumetric oil flow rate depends on the oil inlet temperature, pressure drop between p_i and p_o, rotor speed, as well as the rotating speed of the bearing ring.

The inner oil film has to carry the rotor to balance external forces acting upon it. Therefore, its inner bearing radial clearance c_1 is relatively small to increase the bearing stiffness coefficients. On the contrary, the outer oil film provides the rotor with large damping effects to reduce the rotor deflection at resonances, and suppress the rotordynamic airborne noises of the unbalance whistle and constant tone.

Flow characteristics of the oil films are taken into account in computing the friction power in the bearing (see Fig. 6.6).

The bearing friction power produced in the rotor shaft is given by

$$P_f = F_f U_j = U_j \int_A \tau \, dA = \eta_i \frac{\partial U}{\partial h} A_j U_j \approx \left(\eta_i \frac{U_j}{h} A_j \right) U_j \tag{6.21}$$

where

η_i is the inner oil film dynamic viscosity;
$\partial U / \partial h$ is the oil film velocity gradient;
h is the oil-film thickness;

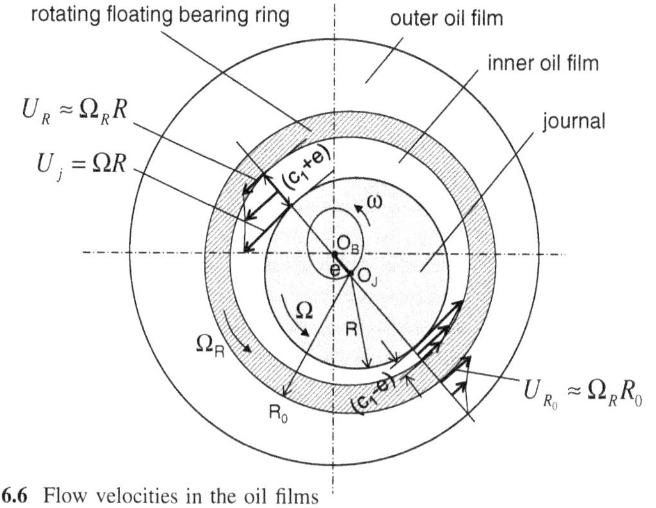

Fig. 6.6 Flow velocities in the oil films

A_j is the journal frictional area;
U_j is the journal circumferential velocity.

In order to reduce the bearing friction power, the velocity gradient is decreased when the bearing ring rotates at a circumferential velocity U_R at the inner bearing clearance. Hence, the bearing friction power is reduced to

$$P_f^* \approx \eta_i \frac{(U_j - U_R)}{h} A_j U_j \tag{6.22}$$

According to Eqs. (6.21) and (6.22), the relative reduction of the bearing friction power is computed.

$$\frac{\Delta P_f}{P_f} = \frac{P_f - P_f^*}{P_f} = 1 - \frac{P_f^*}{P_f} = \frac{U_R}{U_j} \approx \frac{\Omega_R}{\Omega} \equiv \mathrm{RSR} \tag{6.23}$$

where

Ω_R is the ring angular velocity ($\Omega_R < \Omega$);
Ω is the rotor angular velocity.

Equation (6.23) indicates that the friction power reduction is proportional to the ring speed ratio RSR. Therefore, the higher the ring speed ratio, the larger the bearing friction power decreases.

The Petroff's law resulting from the Reynolds lubrication equation computes the torque acting upon the journal in the semi-floating ring bearing.

$$M = R\int_A \tau \, \mathrm{d}A = R\int_A \eta \frac{\partial U}{\partial h} \mathrm{d}A = \frac{2\pi\eta R^3 L\Omega}{c} \tag{6.24}$$

Having applied Eq. (6.24) to the rotating floating ring bearing, one obtains the driving torque acting on the bearing ring inner side,

$$M_i = \frac{2\pi\eta_i R_i^3 L_i (\Omega - \Omega_R)}{c_1} \tag{6.25}$$

and the friction torque on the bearing ring outer side

$$M_o = -\frac{2\pi\eta_o R_o^3 L_o \Omega_R}{c_2} \tag{6.26}$$

Having used the angular momentum law, the ring speed ratio *RSR* is resulted at the steady-state condition ($\ddot{\theta} = 0$).

$$\sum M = M_i + M_o = I_p \ddot{\theta} = 0$$

Thus,

$$\mathrm{RSR}_{\mathrm{steady}} \equiv \frac{\Omega_R}{\Omega} \approx \frac{1}{1 + \frac{\eta_o(T_o)}{\eta_i(T_i)}\left(\frac{L_o}{L_i}\right)\left(\frac{c_1}{c_2}\right)\left(\frac{D_o}{D_i}\right)^3} \tag{6.27}$$

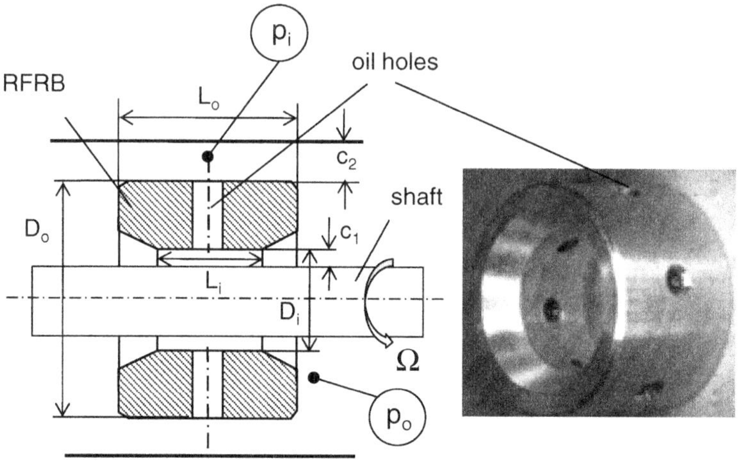

Fig. 6.7 Rotating floating ring bearing geometries

where

η_i, η_o are the dynamic viscosity of the inner and outer oil films;
L_i, L_o are the inner and outer bearing widths (see Fig. 6.7);
c_1, c_2 are the inner and outer radial bearing clearances;
D_i, D_o are the inner and outer bearing diameters;
I_p is the polar mass inertia moment of the rotor.

When the inner and outer relative eccentricities ε_1 and ε_2 are taken into account, the modified steady-state RSR* becomes according to [2].

$$\text{RSR}^*_{\text{steady}} \equiv \frac{\Omega_R^*}{\Omega} \approx \frac{1}{1 + \frac{\eta_o(T_o)}{\eta_i(T_i)}\left(\frac{L_o}{L_i}\right)\left(\frac{c_1\sqrt{1-\varepsilon_1^2}}{c_2\sqrt{1-\varepsilon_2^2}}\right)\left(\frac{D_o}{D_i}\right)^3} \tag{6.28}$$

In the unsteady state at $\ddot{\theta} \neq 0$, the ring speed ratio in Eqs. (6.27) and (6.28) is modified according to [3].

$$\text{RSR}_{\text{unsteady}} = \text{RSR}_{\text{steady}} - \frac{I_p\ddot{\theta}}{2\pi\Omega\left(\frac{\eta_i R_i^3 L_i}{c_1} + \frac{\eta_o R_o^3 L_o}{c_2}\right)} \tag{6.29}$$

Because the inner clearance is much smaller than the outer clearance, the inner oil-film temperature is higher than the outer oil-film temperature at increasing rotor speeds due to the bearing friction. Hence, the oil viscosity ratio between the outer and inner oil films increases (i.e., $\eta_o >> \eta_i$), leading to the ring speed ratio decrease. The computed ring speed ratio versus rotor speed at different oil inlet temperatures of the rotating floating ring bearing is displayed in Fig. 6.8. In the computation, the effective temperatures $T_{b,i}$ and $T_{b,o}$ of the inner and outer oil films

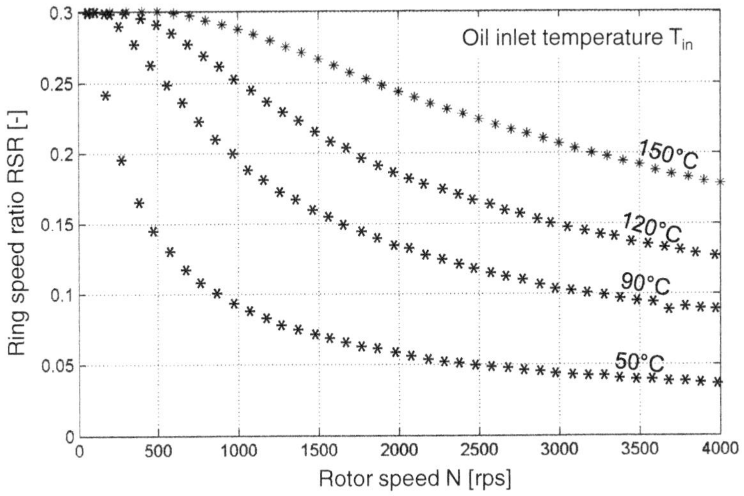

Fig. 6.8 Computed ring speed ratio RSR_{steady} versus rotor speeds

are induced only by the bearing friction without the heat conduction from the bearing center housing to the oil films. In fact, the heat conduction from the bearing center housing increases the outer oil film temperature. Therefore, the ring speed ratio at the turbine side is normally quite larger than the ring speed ratio at the compressor side. The ring speed ratio varies between 15 and 30 % at low rotor speeds to about 90,000 rpm (1,500 cps) at the oil inlet temperature of 90 °C. Hence, the rotor friction power of the rotating floating ring bearing could be reduced by 20–30 % compared to the semi-floating ring bearing in low-end torque (LET), leading to improving the turbocharger's transient behavior.

The higher the rotor speed, the higher the effective inner oil film temperature is, leading to the ring speed reduction. At high inlet oil temperatures, the dynamic viscosity ratio of both outer and inner oil films does not change so much; therefore, the ring speed ratio reduces more slowly and remains nearly constant at higher rotor speeds. At decreasing inlet oil temperatures, the bearing friction increases in the inner oil film, leading to high inner oil film temperatures. As a result, the ring speed is lowered as the inlet oil temperature decreases. Generally, the ring speed ratio decreases with the rotor speed at a constant oil inlet temperature and increases with the oil inlet temperature at a constant rotor speed, as shown in Fig. 6.8. The rotating floating ring bearings have two main advantages: first, the bearing friction reduction, especially in low-end torque; second, the large damping effect due to the large outer oil film. However, they need large oil volumetric flow rate and the rotor orbit is larger than the orbit of the semi-floating ring bearings because of having two oil films.

If we consider the unsteady state in the rotor run-up phase, the ring speed ratio is very low at small rotor speeds and high inlet oil temperatures because the inertia term of the ring speed ratio in Eq. (6.29) is significantly large, especially at high

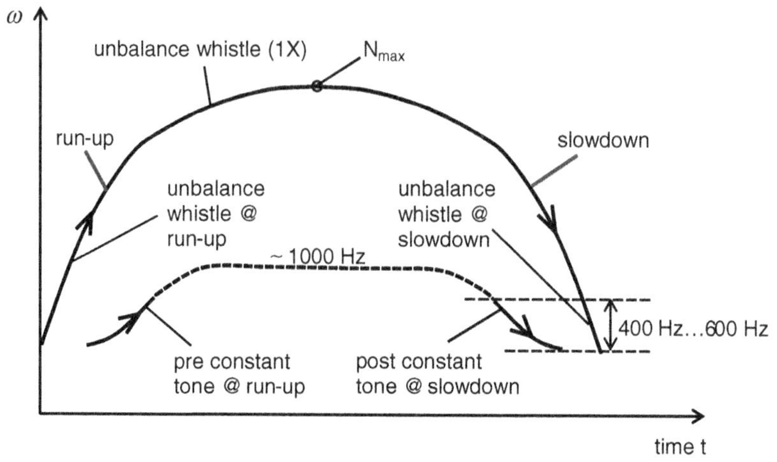

Fig. 6.9 Induced noises in the run-up and slowdown phases

accelerations. At increasing rotor speeds, the inertia term becomes smaller; it leads to the ring speed ratio increase. Then, the ring speed ratio is steady state at further increasing rotor speeds.

6.5 Pre and Post Constant Tone

In the run-up phase, turbocharger rotor is accelerated from the idle condition to maximum rotor speed, as displayed in Fig. 6.9. The oil film temperature in the inner bearing is relatively low and nearly equals the outer oil film temperature at small rotor speeds. Therefore, the ring speed ratio given in Eq. (6.29) increases slowly during the rotor acceleration. According to Eq. (6.20), the inner oil whirl frequency is nearly proportional to the product of the rotor speed and ring speed ratio. As a result, the pre constant tone frequency varies from 400 to 600 Hz in the audible frequency range.

At further increasing rotor speeds to the maximum speed, the inner oil film temperature rises higher than the outer oil film temperature because the viscous friction in the inner oil film is larger than the friction in the outer oil film, leading to viscosity reduction in the inner oil film. Hence, the ring speed ratio reduces with rotor speeds according to Eqs. (6.28) and (6.29). On the one hand, the inner oil whirl frequency decreases with reducing ring speed ratios at high rotor speeds. On the other hand, the inner oil whirl frequency is proportional to the rotor speed, as shown in Eq. (6.20). As a result, the constant tone frequency increases much slowly at high rotor speeds and therefore ranges from 600 to 1,000 Hz. Thus, the constant tone is considered quasi-constant compared to the rotor frequency in the normal operating condition.

In the slowdown phase, turbocharger rotor is decelerated from the maximum rotor speed to the idle condition. During the slowdown, the ring speed ratio does not change too much because the viscosities of both oil films are nearly equal. Therefore, the inner oil whirl frequency decreases from 1,000 to 400 Hz at lowering rotor speeds. Between about 600 and 400 Hz, the post constant tone usually becomes audible in the vehicle cabin, as shown in Fig. 6.9. The low damping effectiveness in the outer oil film is possibly to blame for this audible post constant tone between 600 and 400 Hz. However, it is inaudible at low frequencies (<400 Hz) since the noise level is proportional to frequency squared.

The post constant tone in the slowdown phase is considered more critical than the pre constant tone in the run-up phase because engine noise in the slowdown strongly decreases compared to the run-up. This post constant tone could be reduced by active and passive measures that will be discussed in the following section. However, the post constant tone is sometimes confused with the unbalance whistle and other generated noises of components neighboring the turbocharger because their frequencies also decrease with the rotor speed in the slowdown phase. In this case, other noise sources must be at first improved instead of the post constant tone alone. In fact, such unbalance whistle, fuel-pump noise, and high frequency-order engine noises, etc., could be audible in addition to the post constant tone in some car types. Note that the post constant tone level depends on the vehicle type, and how the turbocharger is mounted in the engine.

6.6 Measures of Reducing Constant Tone Level

As a result, the inner oil whirl amplitude is proportional to the oil whirling volumetric flow rate given in Eq. (6.18); its whirl frequency ω_i depends on the volumetric oil flow rate in the axial direction, rotor and bearing ring speeds, as given in Eqs. (6.19) and (6.20). To reduce the constant tone level, the oil whirling flow rate must be kept as small as possible. There are some possible measures to reduce the constant tone level by minimizing the oil whirl amplitude, as shown in Table 6.1.

6.6.1 Active Measures

The main reason for the oil whirl is the oil volumetric flow rate surplus in the convergent wedge of the inner bearing clearance. To minimize the constant tone level, the oil flow rate surplus must be relieved to keep the inner oil whirl amplitude as small as possible; therefore, the rotor remains in the stable condition. Equation (6.18) shows that the oil whirling flow rate can be minimized by either reducing the oil flow rate surplus itself (the first term) or increasing the oil flow rate in the axial direction (the second term) to compensate the oil flow rate surplus.

Table 6.1 Measures to reduce the constant tone level

Measures		**Positive effects**	**Disadvantages**
Active	Smaller inner bearing clearance $2c_1$	Reducing the amplitude of the constant tone → reduced $e = \varepsilon.c_1$	• High oil temperature → Oil coking • Wear in the mixed lubrication
	Shorter inner bearing width L_i	Reducing the amplitude of the constant tone	• Small inner bearing surface • Wear in the mixed lubrication
	Reduced ring speed ratio *RSR*	Reducing the amplitude of the constant tone	• Increasing bearing friction at LET • Bad transient behavior at LET
Passive	Larger outer bearing clearance $2c_2$	More damping effectiveness in the outer oil film	• Larger rotor deflection → unstable rotor • Bearing modulation → side bands
	Larger bearing outer diameter D_o	More damping effectiveness in the outer oil film	• Increasing blowby and oil leakage due to larger seal rings
	Radial bearing with inner axial grooves	Increasing the axial oil flow rate	• Bearing wear due to oil insufficiency • Bearing modulation → side bands
	Optimizing the turbocharger housing eigenfrequencies	Reducing the resonant amplitudes of the turbocharger housing excited by the constant tone	–
	Optimizing stiffness and damping characteristics of the car frame and periphery components	Reducing the response amplitudes of components excited by the constant tone	–

As a result, the oil whirling flow rate decreases, leading to reducing the constant tone level.

First, some measures are possible to reduce the oil flow rate surplus according to Eq. (6.18). They are namely, reducing the inner bearing clearance c to increase the bearing stiffness coefficient so that the journal eccentricity $e = c.\varepsilon$ becomes smaller; shortening the inner bearing width L_i; reducing the rotor diameter D (=2R); or decreasing the ring speed ratio. However, all these measures should be thoroughly studied before applying them in turbochargers. They could cause some disadvantages, such as the rotor instability, wear due to the bearing inner surface reduction, increasing bearing friction due to mixed lubrication between the journal and bearing, worsening transient behavior in the low end torque due to decreasing the bearing ring speed, and finally oil coking in the inner bearing clearance.

Second, one increases the oil volumetric flow rate in the axial direction by installing a few axial oil grooves in the inner bearing to compensate the oil flow rate surplus. Therefore, the journal whirling amplitude reduces, leading to the constant tone level reduction. However, wear could occur in the bearing because lubricating oil does not sufficiently maintain in the bearing clearance, especially at the start-stop-driving condition or highly unbalanced rotor at the trim balancing.

6.6.2 Passive Measures

The oil film damping effectiveness is defined by the ratio of the dissipated energy in the oil film to the total energy of the vibration system per unit of radian.

The oil film damping effectiveness results in

$$D_{eff} \equiv \frac{1}{2\pi}\left[\frac{E_d}{E_t}\right] = \frac{\pi c \omega x^2}{2\pi(\frac{1}{2}kx^2)} = \frac{c\omega}{k} \tag{6.30}$$

where c is the oil film damping coefficient; k is the rotor-bearing stiffness coefficient;x is the bearing displacement amplitude; ω is the bearing whirl frequency.

The outer bearing clearance is designed large enough to suppress the constant tone level by increasing the damping effectiveness while transmitting through the outer oil film of the bearing. However, an excessively large outer bearing clearance could induce rotor instability because the bearing stiffness coefficient decreases. Another possible measure is the large bearing outer diameter to increase the outer oil film damping and stiffness coefficients; hence, the damping effectiveness of the outer oil film increases in order to damp and suppress the constant tone level. However, the ring speed ratio of the bearing reduces, leading to the bearing friction increase in the rotor and therefore degrading the turbocharger transient behavior at LET. In addition, the turbocharger housing eigenfrequencies could be optimized to reduce its resonance amplitude excited by the constant tone.

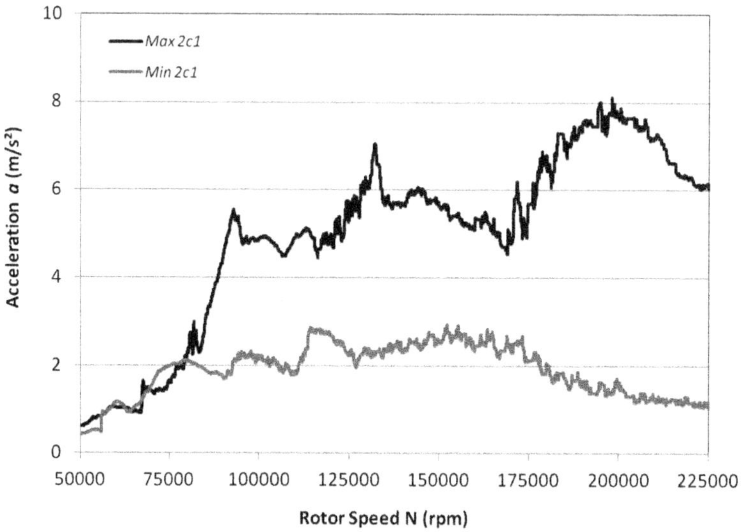

Fig. 6.10 Constant tone levels at min–max inner bearing clearances measured in the bearing housing [4]

Similarly, the stiffness and damping characteristics of the car frame and periphery components neighboring the turbocharger could be improved by shifting their eigenfrequencies to decrease the response amplitudes excited by the constant tone (s. Chapter 7).

In the following section, two exemplary measures of reducing the inner bearing clearance (active measure), and increasing the outer bearing clearance (passive measure) are used to reduce the constant tone level of automotive turbochargers. At the active measure, the diametral inner bearing clearance $2c_1$ is reduced from 15 to 23 μm. As a result, the constant tone level corresponding to the acceleration is decreased from 8 m/s^2 at the maximum bearing clearance of 23 μm to nearly 3 m/s^2 at the minimum bearing clearance of 15 μm, as shown in Fig. 6.10. This result indicates that the smaller the inner bearing clearance, the lower the constant tone level is. The reason is that the small inner bearing clearance increases the bearing stiffness coefficient; therefore, the journal eccentricity becomes smaller. Hence, the oil whirling volumetric flow rate reduces according to Eq. (6.18), leading to the constant tone level reduction. On the contrary, the journal eccentricity increases at larger inner bearing clearances $2c_1$ due to lowered bearing stiffness coefficients, leading to higher constant tone levels. Note that, small inner bearing clearances could cause bearing wear and oil coking in the bearing clearance due to hard particle wear and high oil temperatures induced by the bearing friction, respectively.

The passive measure of increasing the outer diametral bearing clearance $2c_2$ leads to decreasing the constant tone level. Measured results in Fig. 6.11 show that

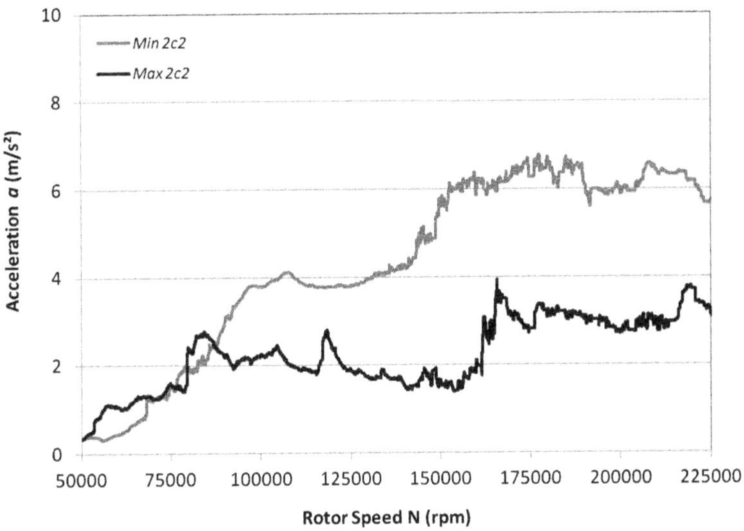

Fig. 6.11 Constant tone levels at min–max outer bearing clearances measured in the bearing housing [4]

the larger the outer oil film thickness, the lower the constant tone level is. The large outer oil film thickness induces less bearing friction; the outer oil film temperature is lowered, leading to increasing the damping effectiveness due to large bearing whirl frequency in the outer oil film. Therefore, the outer oil film damps the constant tone level while transmitting through the outer oil film. The bearing friction in the outer oil film increases with reducing the outer bearing clearance $2c_2$; hence, the oil temperature rises, leading to the oil viscosity reduction in the outer oil film. As a result, the outer oil film damping effectiveness is reduced by decreasing the bearing whirl frequency so that the constant tone level is little damped while passing the outer oil film. As an example, the outer bearing clearance $2c_2$ is increased from 68 to 84 μm; the constant tone level corresponding to the acceleration response is reduced from about 7 m/s^2 at the minimum bearing clearance of 68 μm to nearly 4 m/s^2 at the maximum bearing clearance of 84 μm, as displayed in Fig. 6.11. However, the outer oil film stiffness coefficient reduces by increasing outer bearing clearances. Therefore, the rotor amplitude could become unstable in the outer oil film, leading to bearing wear in the mixed boundary lubrication between the bearing rings and bearing center housing. In this case, it should be carefully proved which maximum outer bearing clearance fulfills the rotordynamic stability condition in the operating speed range.

References

1. Hori, Y.: Hydrodynamic lubrication. Springer, (2006)
2. Muszýnska, A.: Rotordynamics. CRC Press, (2005)
3. Nguyen-Schäfer, H.: Rotordynamics of Automotive Turbochargers. Springer, Berlin-Heidelberg (2012)
4. Nguyen-Schäfer, H., Kleinschmidt, R.: Analysis and nonlinear rotordynamics computation of constant tone in automotive turbochargers, 17th ATK conference. Dresden, Germany (2012)

Chapter 7
Eigenfrequency Modifications to Reduce Constant Tone Level

Constant tone induced by the inner oil whirl is transmitted through the oil films and periphery components neighboring the turbocharger, such as bearing center housing, compressor housing, air filter, charge-air intercooler, exhaust-gas system, and car frame to the vehicle cabin, as shown in Fig. 6.2. While the transmission, the constant tone could excite the periphery components, especially nearby the turbocharger. The resonance amplitudes occur when the component eigenfrequencies $\omega_{c,i}$ and $\omega_{c,k}$ equal the excitation frequencies of the inner oil whirl $\omega_{\mathrm{IOW,min}}$ and $\omega_{\mathrm{IOW,max}}$, as shown in Fig. 7.1.

The response amplitudes of the periphery components additionally amplify the airborne noise level at the bandwidth Δf of the constant tone frequency. The audible noises are usually considered as the constant tone, especially in the run-up and slowdown phases at the frequency range between 400 and 600 Hz, as discussed in Sect. 6.5. Thus, the noise levels strongly depend on the vehicle characteristics; as an example, the constant tone is inaudible in one car type; however, it could be audible in another car type even with the same turbocharger. In this case, the eigenfrequencies of the vehicle types are to blame for the airborne acoustic quality.

The component eigenfrequency is calculated from the component stiffness coefficient K and mass m [1, 4].

$$\omega_c = \sqrt{\frac{K}{m}} \tag{7.1}$$

Figure 7.1 shows the Campbell diagram of a periphery component nearby the turbocharger. Every component eigenfrequency in the synchronous excitation line $1X$ equals the constant tone frequency; therefore, many resonances take place at the cutting points between the component eigenfrequencies and constant tone frequency bandwidth. The periphery component resonance amplitudes intensify the airborne in the constant tone frequency bandwidth, as seen in Fig. 7.1. In this case, the resonance amplitudes of the excited components are also audible in the vehicle cabin, but not alone the constant tone induces the airborne noise.

H. Nguyen-Schäfer, *Aero and Vibroacoustics of Automotive Turbochargers*,
DOI: 10.1007/978-3-642-35070-2_7, © Springer-Verlag Berlin Heidelberg 2013

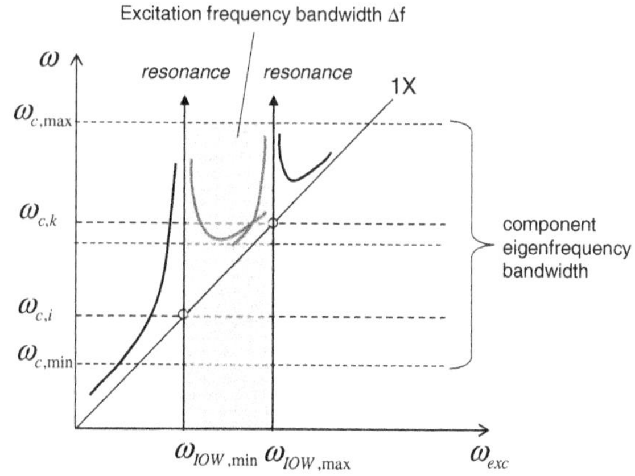

Fig. 7.1 Campbell diagram of periphery components

To reduce or avoid the intensified airborne noise, the component eigenfrequencies should be moved out of the constant tone excitation frequency range so that no resonance takes place in the constant tone frequency bandwidth. According to Eq. (7.1), one can vary the component eigenfrequency by either stiffening the stiffness coefficient or reducing the component mass. As an example, the stiffness coefficient is increased so that the minimum component eigenfrequency exceeds the maximum constant tone frequency, as shown in Fig. 7.2. Thus, the response amplitude is far from the resonance frequency and remains small, leading to lowered airborne noise level. To increase the eigenfrequencies, the periphery components should be tightened and damped in the engine, or their wall thickness is increased. In the case of a long flexible air pipe, it should be fixed at some damping supports between the pipe ends to increase its stiffness coefficient. However, it is not always easy to stiffen all periphery components in the engine.

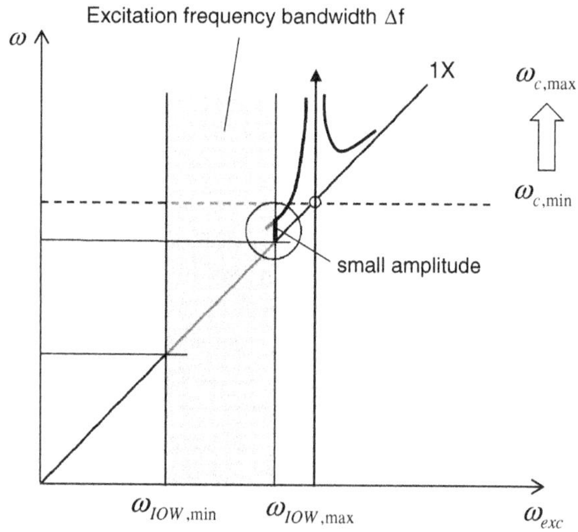

Fig. 7.2 Campbell diagram of optimized periphery components

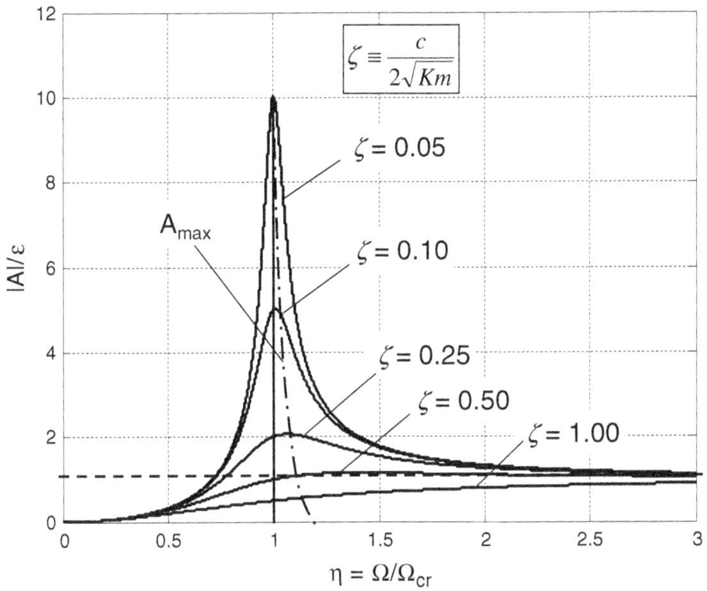

Fig. 7.3 Response amplitudes at various damping ratios ζ

To detect noise of the engine components, a 3-D acoustic camera is used to analyze and visualize the airborne noise field of the engine in different operating conditions. Noise pressure levels are recorded by means of microphone arrays, such as ring-shaped, spherical, or near-field planar arrays with to about 120 microphones. By using the analyzing and evaluating softwares, e.g., Software Noise Image, the noise field-induced by the turbocharger itself and its periphery components are visualized; the noise spectra diagram (Waterfall plot) is additionally displayed. This method is useful to detect which component stiffness should be improved in the applied vehicle type.

Alternatively, the periphery components could be damped in order to reduce the response amplitude $|A|$ even at the resonance $\eta = 1$, as shown in Fig. 7.3. Note that the more the damping ratio is, the lower the vibration amplitude responses.

The dimensionless damping ratio ζ is defined by the ratio between the component stiffness, damping coefficients, and mass [2, 3].

$$\zeta = \frac{c}{2\sqrt{Km}} \tag{7.2}$$

In practice, the excited components are damped by elastomeric materials at the supports in order to reduce the response amplitude. In fact, acoustic engineers have to take the compromise between the periphery component improvement and costs; rotor dynamic engineers have to reduce the induced noise in automotive turbochargers.

References

1. Muszýnska, A.: Rotordynamics. CRC Press, New York (2005)
2. Nguyen-Schäfer, H.: Rotordynamics of automotive turbochargers. Springer, Berlin-Heidelberg (2012)
3. Vance, J.: Rotordynamics of turbomachinery. Wiley, New York (1988)
4. Vance, J., Zeidan, F., Murphy, B.: Machinery vibration and rotordynamics. Wiley, New York (2010)

Conclusions

Besides the functionality of automotive turbochargers, the airborne noise level of turbochargers has recently become a much more important topic, especially in passenger vehicles. To analyze the noise sources of turbochargers, two flow instabilities in the compressor have been studied in this book. First, the airflow instability in the compressor causes rotating stall and surge, which induce the growling and whining noise, respectively. Second, the self-excited instability of the inner oil whirl in the rotating floating ring bearings, leading to the constant tone (howling). In addition to these noise types, the residual unbalance after trim balancing and unbalance changes after a long-term operating period generates the unbalance whistle. All these are the noise sources induced in automotive turbochargers of passenger vehicles. These uncomfortable noise types are transmitted from the noise sources through the turbocharger itself, periphery components nearby the turbocharger, and vehicle frame to the passenger cabin. The noise pressure levels strongly depend on the car types; they are subjectively evaluated by different drivers and passengers for the different car types in various operating conditions.

The noise computational simulation of automotive turbochargers is much more complex and difficult than computational fluid dynamics (CFD) because the noise computational simulation involves many interdisciplinary simulations, such as the noise generation simulation using CFD and computational nonlinear rotordynamics (CNR) in the near-field, acoustic propagation simulation in the periphery components neighboring turbochargers using computational aeroacoustics (CAA) including linearized Euler's equations (LEE) for linear aeroacoustics, and acoustic perturbation equations (APE) using the Lighthill's equation for nonlinear aeroacoustics in the mid-field, and Far-Field simulation of noise radiation into the environment outside the vehicle using the Ffowcs Williams and Hawkings equations (FWHE), which is based on the integral methods using finite element method (FEM) and boundary element method (BEM) for solving the noise pressures of the nonlinear FWHE and Kirchhoff-Helmholtz integral equations. Mostly, the BEM is used to solve the Kirchhoff-Helmholtz integral

H. Nguyen-Schäfer, *Aero and Vibroacoustics of Automotive Turbochargers*,
DOI: 10.1007/978-3-642-35070-2, © Springer-Verlag Berlin Heidelberg 2013

equation with aero- and rotordynamic noise sources by summation of the induced noise pressures over the boundary surface for three-dimensional acoustic field problems. Therefore, it is a big challenge in the near future to develop a commercial computational code for nonlinear aeroacoustics, fluid dynamics, and nonlinear rotordynamic simulations of automotive turbochargers in which the periphery components nearby turbochargers are taken into account in passenger vehicles.

As a matter of fact, the acoustic evaluation usually comes at the end of the product development phase because the new vehicle model has never existed long before its launching on the market. Therefore, airborne noise could not be evaluated in the hot gas test rig in advance since the acoustic quality in the vehicle cabin depends on the different influences of the vehicle frame and periphery components nearby turbochargers. Sometimes, modified radial bearing types must be applied in a certain new car model although they have not yet been thoroughly evaluated for durability and wear in a long-term operating period. For this reason, the airborne noise should be deeply understood so that one can react and improve much more quickly at the end of the development phase, just shortly before the start of production (SOP) for every customer's vehicle model.

Further Reading

1. Bartel, D.: Simulation von Tribosystemen. Springer-Teubner, German (2010)
2. Cremer, L., Möser, M.: Technische Akustik (in German), 5. Auflage, Springer, Berlin (2003)
3. Crocker, M.J.: Noise and vibration control. Wiley, Inc. (2007)
4. Cumpsty, N.A.: Compressor aerodynamics. Krieger Publishing Company, New York (2004)
5. Fahy, F., Gardonio, P.: Sound and structural vibration, 2nd edn. Academic, New York (2007)
6. Greitzer, E.M.: The stability of pumping systems, the 1980 Freeman scholar lecture. J. Fluids Eng. **103** (1981)
7. Hori, Y.: Hydrodynamic lubrication. Springer, Berlin (2006)
8. Muszýnska, A.: Rotordynamics. CRC Press, Boca Raton (2005)
9. Nguyen-Schäfer, H.: Rotordynamics of automotive turbochargers. Springer, Heidelberg (2012)
10. Nguyen-Schäfer, H., Kleinschmidt, R.: Analysis and nonlinear rotordynamics computation of constant tone in automotive turbochargers. In: Proceedings of the 17th ATK Conference, Dresden, Germany (2012)
11. Nguyen-Schäfer, H., Sprafke, P.: Numerical study on interaction effects of the bubbles induced by air-release and cavitation in hydraulic systems, tenth bath international fluid power workshop, Bath, UK. Research Studies Press Ltd., Hertfordshire, England (1997)
12. Nguyen-Schäfer, H.: Nonlinear rotordynamics computations of automotive turbochargers using rotating floating ring bearings at high rotor speeds. In: Proceedings of the 10th SIRM International Conference, Berlin, Germany (2013)
13. Nordmann, R.: Schwingungsberechnung von nichtkonservativen rotoren mit hilfe von links—und rechts-eigenvektoren (in German), vol. 269. VDI Bericht (1976)
14. Rieger, N.F.: Rotordynamics 2—problems in turbomachinery, CISM courses and lectures No. 297 (1988)
15. Rieger, N.F.: Balancing of rigid and flexible rotors. U.S. DoD (1986)
16. Schmied, J.: Program MADYN 2000 for computation of rotordynamics, Delta J.S., Zurich (2011)
17. Schobeiri, M.: Turbomachinery flow physics and dynamic performance, 2nd edn, Springer, Berlin (2012)
18. Vance, J.: Rotordynamics of turbomachinery. Wiley Inc., New York (1988)
19. Vance, J., Zeidan, F., Murphy, B.: Machinery Vibration and Rotordynamics. Wiley, New York (2010)
20. Whitfield, A., Baines, N.C.: Design of radial turbomachines, Pearson education, Longman scientific and technical (1990)
21. Aris, R.: Vectors, tensors, and the basic equations of fluid mechanics. Dover Publications Inc., New York (1989)
22. Synge, J.L., Child, A.: Tensor calculus. Dover Publications Inc., New York (1969)

H. Nguyen-Schäfer, *Aero and Vibroacoustics of Automotive Turbochargers*,
DOI: 10.1007/978-3-642-35070-2, © Springer-Verlag Berlin Heidelberg 2013

Appendix A
Thermodynamic Characteristics of Perfect Gases

Some essential thermodynamic characteristics of gases are required to be known in turbocharging. They have usually been applied to the turbocharging of downsized engines where the charge air and exhaust gas are assumed as perfect gases.

Total temperature T_t in Kelvin (K) is resulted from the sum of the static T_s (K) and dynamic temperatures T_{dyn} (K). The static temperature is measured at the wall where the gas velocity equals zero due to the viscous boundary layer.

$$\begin{aligned} T_t &= T_s + T_{\text{dyn}} \\ &= T_s + \frac{v^2}{2c_p} \end{aligned} \tag{A.1}$$

where

v is the gas velocity;

c_p is the heat capacity at constant pressure.

Total pressure p_t is calculated from the isentropic gas equation of

$$\begin{aligned} p_t &= p_s \left(\frac{T_t}{T_s}\right)^{\frac{\kappa}{\kappa-1}} \\ &= p_s \left(1 + \frac{\kappa-1}{2} M^2\right)^{\frac{\kappa}{\kappa-1}} \end{aligned} \tag{A.2}$$

where

p_s is the static pressure;

$\kappa = c_p/c_v$ is the isentropic exponent of gas;

c_v is the heat capacity at constant volume;

$M = v/c$ is the Mach number of gas;

c = the sound speed is calculated with the gas constant R_a.

$$c = \sqrt{\kappa R_a T_s} \tag{A.3}$$

Specific total enthalpy h_t is resulted from the sum of the gas specific enthalpy and specific kinetic energy of gas.

H. Nguyen-Schäfer, *Aero and Vibroacoustics of Automotive Turbochargers*,
DOI: 10.1007/978-3-642-35070-2, © Springer-Verlag Berlin Heidelberg 2013

$$h_t(T) = h(T) + \frac{v^2}{2} \tag{A.4}$$

The gas specific enthalpy h (enthalpy per mass unit, J/kg) is defined by

$$\begin{aligned} h(T) &= c_p(T - T_0) \\ &= u(T) + \frac{p}{\rho} \\ &= u(T) + \mathrm{R}_a T \end{aligned} \tag{A.5}$$

within

ρ is the gas density;
T_0 is the reference temperature;
u is the gas specific internal energy.

The gas specific internal energy equals the product of the heat capacity at constant volume c_v and temperature difference of gas ΔT.

$$u(T) = c_v(T - T_0) - \mathrm{R}_a T_0 = c_v \Delta T - \mathrm{R}_a T_0 \tag{A.6}$$

where the gas enthalpy at $T_0 = 273.14$ K (0 °C) is defined by

$$h(T_0) \equiv 0 \tag{A.7}$$

Thus, one obtains the gas-specific internal energy according to Eq. (A.7)

$$u(T_0) \equiv -\mathrm{R}_a T_0 \tag{A.8}$$

Appendix B
Transformation of Coordinates

The discretized noise pressure equations in the local orthogonal coordinate system (ξ_1, ξ_2, ξ_3) of the finite element grid must be transformed into the global Cartesian coordinate system (x_1, x_2, x_3) in the FEM computation, as shown in Fig. B.1. To carry out this transformation, some necessary mathematical backgrounds are briefly described. More details can be found in many literatures [17, 21, 22].

The position vector r in the control volume V is written in

- the Cartesian coordinate system (x_1, x_2, x_3)

$$\boldsymbol{r} = x_1\boldsymbol{e}_1 + x_2\boldsymbol{e}_2 + x_3\boldsymbol{e}_3. \tag{B.1}$$

- the curvilinear coordinate system (ξ_1, ξ_2, ξ_3)

$$\boldsymbol{r} = \xi_1\boldsymbol{g}_1 + \xi_2\boldsymbol{g}_2 + \xi_3\boldsymbol{g}_3. \tag{B.2}$$

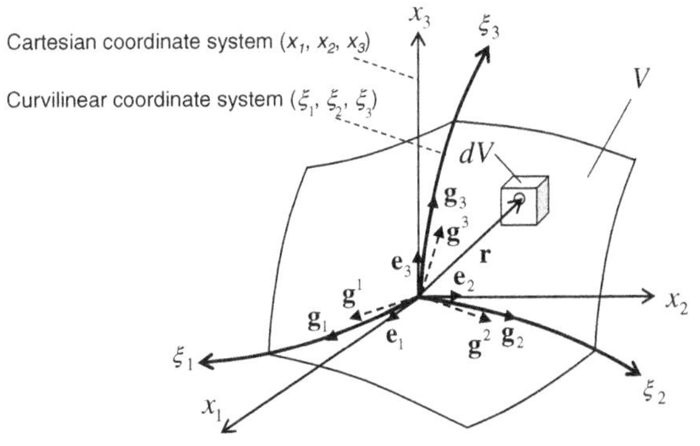

Fig. B.1 Transformation of coordinates

H. Nguyen-Schäfer, *Aero and Vibroacoustics of Automotive Turbochargers*,
DOI: 10.1007/978-3-642-35070-2,

where $\boldsymbol{e}_1$, $\boldsymbol{e}_2$, and $\boldsymbol{e}_3$ are the covariant basis vectors (unit vectors) of the Cartesian coordinate system;

$\boldsymbol{g}_1$, $\boldsymbol{g}_2$, and $\boldsymbol{g}_3$ are the covariant basis vectors (non-unit vectors) of the curvilinear coordinate system;

x_i is the covariant vector component of **r** for $i = 1, 2, 3$;

ξ_i is the covariant vector component of **r** for $i = 1, 2, 3$.

The covariant basis vectors tangent to the curvilinear coordinate are defined by using Einstein summation symbol.

$$\begin{aligned} \boldsymbol{g}_i &= \frac{\partial \boldsymbol{r}}{\partial \xi_i} = \frac{\partial (x_j \boldsymbol{e}_j)}{\partial \xi_i} \\ &\equiv \frac{\partial x_j}{\partial \xi_i} \boldsymbol{e}_j = \sum_{j=1}^{3} \frac{\partial x_j}{\partial \xi_i} \boldsymbol{e}_j \end{aligned} \tag{B.3}$$

The contravariant basis vectors normal to the surfaces (ξ_j, ξ_k) of the curvilinear coordinate system are given by

$$\boldsymbol{g}^j \equiv \frac{\partial \xi_j}{\partial x_m} \boldsymbol{e}_m = \sum_{m=1}^{3} \frac{\partial \xi_j}{\partial x_m} \boldsymbol{e}_m \tag{B.4}$$

The scalar product of the co- and contravariant basis vectors is given by

$$\boldsymbol{g}_k \boldsymbol{g}^j \equiv \frac{\partial x_i}{\partial \xi_k} \cdot \frac{\partial \xi_j}{\partial x_m} \boldsymbol{e}_i \boldsymbol{e}_m = \frac{\partial \xi_j}{\partial \xi_k} \equiv \delta_k^j \tag{B.5}$$

where $\delta_k{}^j$ is the Kronecker delta; $\delta_k{}^j = 0$ if $j \neq k$; $\delta_k{}^j = 1$ if $j = k$.

The metric coefficients of the curvilinear coordinate system are defined by

- Covariant metric coefficient:

$$g_{ij} = \boldsymbol{g}_i \cdot \boldsymbol{g}_j \tag{B.6}$$

- Contravariant metric coefficient:

$$g^{ij} = \boldsymbol{g}^i \cdot \boldsymbol{g}^j \tag{B.7}$$

- Mixed metric coefficient:

$$g_i^j = \boldsymbol{g}_i \cdot \boldsymbol{g}^j \tag{B.8}$$

A fluid vector $\boldsymbol{v}$ can be formulated in the curvilinear coordinate system

$$\begin{aligned} \boldsymbol{v} &= v^1\boldsymbol{g}_1 + v^2\boldsymbol{g}_2 + v^3\boldsymbol{g}_3 \equiv v^i\boldsymbol{g}_i(\text{covariant basis vectors}) \\ &= v_1\boldsymbol{g}^1 + v_2\boldsymbol{g}^2 + v_3\boldsymbol{g}^3 \equiv v_i\boldsymbol{g}^i(\text{contravariant basis vectors}) \end{aligned} \tag{B.9}$$

where

v^1, v^2, v^3 are the contravariant vector components (raised indexes);

v_1, v_2, v_3 are the covariant vector components (lowered indexes).

The Jacobian determinant is applied to transform the coordinates and are given by

$$\boldsymbol{J} \equiv \begin{vmatrix} \frac{\partial x_1}{\partial \xi_1} & \frac{\partial x_1}{\partial \xi_2} & \frac{\partial x_1}{\partial \xi_3} \\ \frac{\partial x_2}{\partial \xi_1} & \frac{\partial x_2}{\partial \xi_2} & \frac{\partial x_2}{\partial \xi_3} \\ \frac{\partial x_3}{\partial \xi_1} & \frac{\partial x_3}{\partial \xi_2} & \frac{\partial x_3}{\partial \xi_3} \end{vmatrix} = \varepsilon_{ijk}\left(\frac{\partial x_i}{\partial \xi_1}\right)\left(\frac{\partial x_j}{\partial \xi_2}\right)\left(\frac{\partial x_k}{\partial \xi_3}\right); \ \forall i, j, k = 1, 2, 3. \tag{B.10}$$

within $\varepsilon_{\mathrm{ijk}}$ is the permutation symbol defined by

$$\varepsilon_{\mathrm{ijk}} \equiv \begin{cases} 0 & \text{if } i = j \text{ or } j = k, \text{ or } k = i \\ 1 & \text{if ijk is an even permutation} (= 123) \\ -1 & \text{if ijk is an odd permutation} (= 132) \end{cases} \tag{B.11}$$

In an orthogonal basis, the Jacobian determinant is resulted from Eqs. (B.3), (B.6), and (B.10).

$$\boldsymbol{J} = \sqrt{g_{11} \cdot g_{22} \cdot g_{33}} \tag{B.12}$$

Using Eq. (B.10), the infinitesimal volume dV is computed in the curvilinear coordinate system.

$$\begin{aligned} dV &\equiv dx_1 dx_2 dx_3 \\ &= d\boldsymbol{X}^1(d\boldsymbol{X}^2 \times d\boldsymbol{X}^3) \\ &= \varepsilon_{ijk}\left(\frac{\partial x_i}{\partial \xi_1}d\xi_1\right)\cdot\left(\frac{\partial x_j}{\partial \xi_2}d\xi_2\right)\cdot\left(\frac{\partial x_k}{\partial \xi_3}d\xi_3\right) \\ &= \boldsymbol{J}.(d\xi_1 d\xi_2 d\xi_3) \end{aligned} \tag{B.13}$$

where the vector $d\boldsymbol{X}^{\mathrm{i}}$ is defined by

$$d\boldsymbol{X}^{\mathrm{i}} = \left[\frac{\partial x_i}{\partial \xi_1} \ \frac{\partial x_i}{\partial \xi_2} \ \frac{\partial x_i}{\partial \xi_3}\right]^T, \ \forall \mathrm{i} = 1, 2, 3. \tag{B.14}$$

The velocity $\boldsymbol{v}$ does not change at changing the coordinate system. Therefore, the volume integral of the velocity $\boldsymbol{v}$ depending on $x_i = x_i\ (\xi_1, \xi_2, \xi_3)$ for $i = 1, 2, 3$ over the control volume V is resulted from Eq. (B.13).

$$\int_V \boldsymbol{v}(x_1, x_2, x_3)dV = \int_V \boldsymbol{v}(\xi_1, \xi_2, \xi_3)\mathbf{J}.d\xi_1 d\xi_2 d\xi_3 \tag{B.15}$$

Similarly, the surface integral of a function $\varphi(\boldsymbol{X})$ is calculated over the tangential surface S_T in the curvilinear coordinate system, where $\boldsymbol{X}(\lambda_1, \lambda_2)$

parameterizes the surface S in the Cartesian coordinate system.

$$\int_S \varphi(\boldsymbol{X})\boldsymbol{n}.\mathrm{d}S = \int_{S_T} \varphi[\boldsymbol{X}(\lambda_1, \lambda_2)].\left(\frac{\partial \boldsymbol{X}}{\partial \lambda_1} \times \frac{\partial \boldsymbol{X}}{\partial \lambda_2}\right)\mathrm{d}\lambda_1 \mathrm{d}\lambda_2 \tag{B.16}$$

The coordinate transformation gives the operators written in Einstein summation convention in the curvilinear coordinate system:

- Gradient operator of a scalar function of $\varphi = \varphi(\xi_1, \xi_2, \xi_3)$

$$\nabla\varphi = \frac{\partial \phi}{\partial \xi_i}\boldsymbol{g}^i \equiv \varphi_{,i}\boldsymbol{g}^i \tag{B.17}$$

- Divergence operator of a vector $\boldsymbol{v} = \boldsymbol{v}(v^1, v^2, v^3)$

$$\begin{aligned}\nabla.\boldsymbol{v} &= \frac{\partial}{\partial \xi_i}(\boldsymbol{g}_j v^j)\boldsymbol{g}^i = \left[\left(\frac{\partial \boldsymbol{g}_j}{\partial \xi_i}\right)v^j + \left(\frac{\partial v^j}{\partial \xi_i}\right)\boldsymbol{g}_j\right]\boldsymbol{g}^i \\ &= v^j_{,i} + v^k \Gamma^j_{ik}\end{aligned} \tag{B.18}$$

where the first term on the rhs of Eq. (B.18) defines the derivative of v^j with respect to ξ_i.

$$v^j_{,i} \equiv \frac{\partial v^j}{\partial \xi_i} \tag{B.19}$$

The Christoffel symbol of the second kind is defined by

$$\Gamma^j_{ik} \equiv \frac{\partial^2 x_m}{\partial \xi_i \partial \xi_k} \cdot \frac{\partial \xi_j}{\partial x_m} \tag{B.20}$$

- Curl operator of a vector $\boldsymbol{v} = \boldsymbol{v}(v_1, v_2, v_3)$

$$\begin{aligned}\nabla \times \boldsymbol{v} &= \boldsymbol{g}^k \frac{\partial}{\partial \xi_k} \times (\boldsymbol{g}^i v_i) = \boldsymbol{g}^k \times \boldsymbol{g}^i v_{i,k} + \boldsymbol{g}^k \times \boldsymbol{g}^i_{,k} v_k \\ &= \varepsilon^{kim}\boldsymbol{g}_m v_{i,k}\end{aligned} \tag{B.21}$$

where ε^{kim} is the permutation symbol given in Eq. (B.11);

The third term on the rhs of Eq. (B.21) defines the derivative of v_i with respect to ξ_k.

$$v_{i,k} \equiv \frac{\partial v_i}{\partial \xi_k} \tag{B.22}$$

Appendix C
Solutions of the Characteristic Equation with Complex Coefficients

The characteristic quadratic equation with complex coefficients is derived from the vibration equation.

$$D(s) \equiv s^2 + (a + \mathrm{jb})s + (c + \mathrm{jd}) = 0 \tag{C.1}$$

where

$s = (\alpha \pm \mathrm{j}\omega_n) \in \mathbf{C}$ are the complex eigenvalues;
$a, b, c, d \in \mathbf{R}$ are the real numbers.

By solving the characteristic equation $D(s) = 0$, the eigenvalue s results in

$$s_{1,2} = -\left(\frac{a + \mathrm{jb}}{2}\right) \pm \sqrt{\left(\frac{a + \mathrm{jb}}{2}\right)^2 - (c + \mathrm{jd})} = \alpha \pm \mathrm{j}\omega_n \tag{C.2}$$

After calculating and rearranging the real and imaginary terms of the eigenvalue, one obtains [8]

$$\alpha = -\frac{a}{2} \pm \frac{1}{\sqrt{2}} \left[\left(\frac{a^2 - b^2}{4} - c \right) + \sqrt{\left(\frac{a^2 - b^2}{4} - c \right)^2 + \left(\frac{\mathrm{ab}}{2} - d \right)^2} \right]^{\frac{1}{2}} \tag{C.3}$$

and

$$\omega_n = \frac{b}{2} \pm \frac{1}{\sqrt{2}} \left[-\left(\frac{a^2 - b^2}{4} - c \right) + \sqrt{\left(\frac{a^2 - b^2}{4} - c \right)^2 + \left(\frac{\mathrm{ab}}{2} - d \right)^2} \right]^{\frac{1}{2}} \tag{C.4}$$

Necessary condition for the rotordynamic stability is that the real term α of the eigenvalue must be negative.

H. Nguyen-Schäfer, *Aero and Vibroacoustics of Automotive Turbochargers*,
DOI: 10.1007/978-3-642-35070-2,

$$\alpha = -\frac{a}{2} \pm \frac{1}{\sqrt{2}} \left[-E + \sqrt{E^2 + F^2} \right]^{\frac{1}{2}} < 0 \tag{C.5}$$

within

$$E \equiv -\left(\frac{a^2 - b^2}{4} - c \right) = -\left(\frac{a^2 - b^2}{4} \right) + c \tag{C.6}$$

$$F \equiv \left(\frac{\text{ab}}{2} - d \right) \tag{C.7}$$

Thus, the stability condition ($\alpha < 0$) for the rotor given in Eq. (C.5) becomes after a few calculation steps.

$$a\,(ac + bd) - d^2 \geq 0 \tag{C.8}$$

Appendix D
Stability Condition of Linear Aeroacoustics

The eigenvalues analysis is based on the free vibration response of the linear vibration. The free vibration response is resulted by superimposing the harmonic components of the linear aeroacoustic system without excitation of noise source. In fact, the free noise vibration response is exactly the homogeneous solutions of the vibration equations. Each harmonic component has a complex eigenvalue in which the imaginary part is the eigenfrequency (natural frequency) of the vibration system; the real part indicates the stability characteristic of the vibration system.

The free noise vibration response of the system is written in

$$r(t) = \sum_{i=1}^{N} r_i e^{\lambda_i t} \tag{D.1}$$

where

r_i is the response amplitude corresponding to the eigenvalue;
λ_i is the complex eigenvalue of the eigenmode i;
N is the number of degrees of freedom (DOF) of the vibration system.

The complex eigenvalue consists of the real and imaginary parts.

$$\lambda = \alpha \pm \mathrm{j}\omega_d \tag{D.2}$$

within α is the real part of the eigenvalue, called the growth/decay rate; ω_d is the imaginary part of the complex eigenvalue (eigenfrequency or damped natural frequency).

Substituting λ into Eq. (D.1), one obtains the vibration free response by applying Euler's theorem.

H. Nguyen-Schäfer, *Aero and Vibroacoustics of Automotive Turbochargers*,
DOI: 10.1007/978-3-642-35070-2,

$$
\begin{aligned}
r(t) &= \sum_{i=1}^{N} r_{i,fw} e^{\alpha_i t} e^{j\omega_i t} + \sum_{i=1}^{N} r_{i,bw} e^{\alpha_i t} e^{-j\omega_i t} \\
&= \sum_{i=1}^{N} r_{i,fw} e^{\alpha_i t} [\cos(\omega_i t) + j\sin(\omega_i t)] + \sum_{i=1}^{N} r_{i,bw} e^{\alpha_i t} [\cos(-\omega_i t) + j\sin(-\omega_i t)] \\
&\equiv x(t) + jy(t)
\end{aligned}
\tag{D.3}
$$

The parts in the square brackets [ ] in Eq. (D.3) are the harmonic vibrations with the eigenfrequencies of the forward and backward frequency components. Hence, the maximal amplitude is limited at 1 at any time t. On the contrary, the terms in front of the square brackets [ ] containing an exponential function exp $(\alpha_i t)$ change with time. The amplitude of the noise response depends on the growth/decay rate α_i. If the rate α_i is positive, the noise amplitude increases with time without limit. The noise response is unstable at any positive rate α; therefore, α is the key parameter for the stability analysis of the noise in the linear aeroacoustics.

Cases of the growth/decay rate α for the noise response behavior:

- $\alpha < 0$: stability;
- $\alpha = 0$: onset of instability (threshold of instability);
- $\alpha > 0$: instability.

At a decay rate ($\alpha < 0$), the vibration amplitude with the frequency ω diminishes in a short time. The envelope of the vibration is the exponential function r_0.exp (αt) that begins from the initial position r_0 at time $t = 0$ and decreases with time to zero. The corresponding forward orbit decays from the initial position r_0 to zero as $t \to \infty$.

On the contrary, at a growth rate ($\alpha > 0$), the vibration begins at the initial position and increases with time exponentially; therefore, the vibration is unstable.

At the rate of $\alpha = 0$, the vibration amplitude is unchanged with time, neither decreasing nor increasing with time because the amplitude of the function r_0.exp (αt) always remains constant at r_0 at any time t. The noise vibration moves in a circular orbit that is called limit cycle, which occurs in the Hopf bifurcation. The vibration is at the threshold of instability (onset of instability).

To study noise vibration stability, the complex eigenvalues λ are analyzed in the λ plane, as shown in Fig. D.1. The complex eigenvalue and its conjugated eigenvalue have the common real part and opposite imaginary part with positive and negative signs that correspond to forward and backward damped natural frequencies. If both these eigenvalues lie in the half-left plane ($\alpha < 0$), the noise vibration behavior is stable; on the half-right plane ($\alpha > 0$), it is unstable. In case of the threshold of instability ($\alpha = 0$), the noise vibration moves with the damped natural frequency ω_d at the constant amplitude, neither decreasing nor increasing with time.

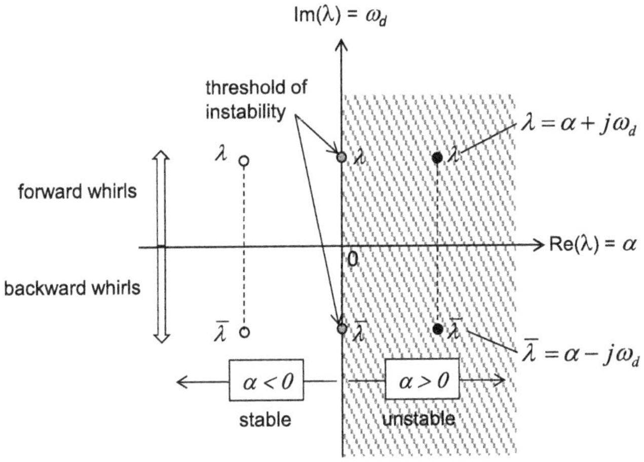

Fig. D.1 Stability analysis in the λ plane

In order to maintain the noise vibration in a stable condition, all eigenvalues of the vibration system must lie on the left half-plane, so that the growth/decay rates are always negative. In this case, there are two possibilities, as shown in Fig. D.1:

- When they locate in the upper left half-plane, the vibration response is a stable forward whirl and decays with time from the initial position to zero.
- When they locate in the lower left half-plane, the vibration response is a stable backward whirl and decays with time from the initial position to zero, but the whirl direction is opposite to the forward whirl.

Appendix E
Routh-Hurwitz Stability Criterion

Characteristic equation is a polynomial equation of nth order.

$$D(\lambda) = a_n\lambda^n + a_{n-1}\lambda^{n-1} + \ldots + a_2\lambda^2 + a_1\lambda + a_0 = 0 \tag{E.1}$$

where

λ are the complex eigenvalues with $\lambda = \alpha + j\omega_d$;
a_i is the equation coefficients, $i = 0, 1, 2, \ldots, n$;
n is the polynomial exponent that equals twice DOF ($n = 2\ N$).

Having used the Gauss theorem, Eq. (E.1) can be written in the polynomial with the n order of λ where λ_i are the roots of $D(\lambda)$.

$$D(\lambda) = a_n(\lambda - \lambda_1)(\lambda - \lambda_2)\cdots(\lambda - \lambda_{n-1})(\lambda - \lambda_n) \tag{E.2}$$

Finding the roots λ_i of the polynomial $D(\lambda)$ in Eq. (E.1) is quite difficult, especially at a large number of DOFs. In fact, it need not be known the exact values of the roots λ_i, but only the real parts α of the eigenvalues are necessary for the stability analysis. In this case, the Routh-Hurwitz stability criterion is applied to analyze the eigenvalues in linear vibration system.

According to the Routh-Hurwitz criterion, the vibration response is stable when the following two conditions are satisfied:

1. All coefficients a_i in the characteristic equation (E.1) must be nonzero and have the same positive or negative signs (necessary condition).
2. Hurwitz determinant D_{n-1}, and its diagonal determinants D_i for $i = 1, 2, \ldots, n-2$ must be positive (sufficient condition).

H. Nguyen-Schäfer, *Aero and Vibroacoustics of Automotive Turbochargers*,
DOI: 10.1007/978-3-642-35070-2, © Springer-Verlag Berlin Heidelberg 2013

$$D_{n-1} = \begin{vmatrix} a_{n-1} & a_{n-3} & a_{n-5} & . & \ldots & . & 0 \\ a_n & a_{n-2} & a_{n-4} & . & \ldots & . & 0 \\ 0 & a_{n-1} & a_{n-3} & a_{n-5} & \ldots & . & . \\ 0 & a_n & a_{n-2} & a_{n-4} & \ldots & . & . \\ 0 & 0 & . & . & \ldots & a_1 & 0 \\ 0 & 0 & . & . & \ldots & a_2 & a_0 \\ 0 & 0 & . & . & a_5 & a_3 & a_1 \end{vmatrix} > 0 \tag{E.3}$$

where the coefficients a_{n-i} equal zero at all negative indices $(n-i)$ for $i = 0$, $1,\ldots, n$.

The Hurwitz determinant D_{n-1} and diagonal determinants D_i for the characteristic equation $D(\lambda)$ of n order ($n = 2$ DOF) can be written in

$$D_1 = |a_{n-1}| > 0;$$

$$D_2 = \begin{vmatrix} a_{n-1} & a_{n-3} \\ a_n & a_{n-2} \end{vmatrix} > 0;$$

$$D_3 = \begin{vmatrix} a_{n-1} & a_{n-3} & a_{n-5} \\ a_n & a_{n-2} & a_{n-4} \\ 0 & a_{n-1} & a_{n-3} \end{vmatrix} > 0\ldots$$

$$D_{n-2} = \begin{vmatrix} a_{n-1} & a_{n-3} & a_{n-5} & \ldots & . & 0 \\ a_n & a_{n-2} & a_{n-4} & \ldots & . & 0 \\ 0 & a_{n-1} & a_{n-3} & \ldots & . & 0 \\ 0 & a_n & a_{n-2} & \ldots & . & a_0 \\ 0 & 0 & . & \ldots & a_3 & a_1 \\ 0 & 0 & . & \ldots & a_4 & a_2 \end{vmatrix} > 0.$$

About the Author

Dr. Hung Nguyen-Schäfer is a senior expert in development of turbochargers at Bosch Mahle Turbo Systems (BMTS) in Germany. He received his B.Sc. and M.Sc. in Mechanical Engineering focusing on nonlinear vibrations in fluid mechanics from the University of Karlsruhe, Germany in 1985 and his Ph.D. in nonlinear thermo- and fluid dynamics from the same university. In 1988 he joined the Bosch company and worked as a technical manager on many development projects.

Since 2007, he has been in charge of rotordynamics, bearing, and designing platforms of automotive turbochargers at Bosch Mahle Turbo Systems (BMTS) located in Stuttgart, Germany, the joint venture of Bosch and Mahle. He is also the author of *Rotordynamics of Automotive Turbochargers*, Springer, 2012.

H. Nguyen-Schäfer, *Aero and Vibroacoustics of Automotive Turbochargers*,
DOI: 10.1007/978-3-642-35070-2, © Springer-Verlag Berlin Heidelberg 2013

Index

H. Nguyen-Schäfer, *Aero and Vibroacoustics of Automotive Turbochargers*,
DOI: 10.1007/978-3-642-35070-2,

CPSIA information can be obtained
at www.ICGtesting.com
Printed in the USA
BVOW07*1844150816
459086BV00001B/1/P

9 783642 350696